Workshop Processes, Practices and Materials

Second edition

Bruce J. Black, C.Eng., MIEE

Formerly workshop director
(wood, metal, plastics)
Gwent College of Higher Education

ARNOLD

A member of the Hodder Headline Group
LONDON • SYDNEY • AUCKLAND

First published in Great Britain in 1979
Twelfth impression 1996

Second edition published in 1997 by
Arnold, a member of the Hodder Headline Group,
338 Euston Road, London NW1 3BH

British Library Cataloguing in Publication Data
A catalogue record for this book is available from the British Library

ISBN 0 340 69252 9

Typeset in 10/12 pt Times by Photoprint, Torquay, Devon
Printed and bound in Great Britain by St Edmundsbury Press, Bury St Edmunds, Suffolk
and J W Arrowsmith Ltd, Bristol

To my wife Gillian and children Susan and Andrew

Contents

Preface   xi

Acknowledgements   xii

1 Safe practices   1
1.1 Health and Safety at Work Act 1974 (HSW Act)   1.2 Health and
safety organisation   1.3 Employers' responsibilities   1.4 Safety
policy   1.5 Safety Representatives and Safety Committees Regulations
1977   1.6 Employees' responsibilities   1.7 New regulations for health
and safety at work   1.8 Management of Health and Safety at Work
Regulations 1992   1.9 Provision and Use of Work Equipment
Regulations 1992 (PUWER)   1.10 Workplace (Health, Safety and
Welfare) Regulations 1992   1.11 Personal Protective Equipment at
Work Regulations 1992   1.12 The Manual Handling Operations
Regulations 1992   1.13 Good handling techniques   1.14 The
Reporting of Injuries, Diseases and Dangerous Occurrences Regulations
1985 (RIDDOR)   1.15 Noise at Work Regulations 1989
1.16 Electrical hazards   1.17 Safety signs and colours   1.18 Fire
1.19 Causes of accidents   1.20 General health and safety precautions

2 Hand processes   30
2.1 Engineer's files   2.2 The hacksaw   2.3 Cold chisels   2.4 Scrapers
2.5 Engineer's hammers   2.6 Screwdrivers   2.7 Taps   2.8 Dies
2.9 Powered hand tools

3 Marking out   44
3.1 Datum   3.2 Co-ordinates   3.3 Marking out equipment
3.4 Examples of marking out

4 Sheet-metal operations   61
4.1 Cutting and bending sheet metal   4.2 Development

5 Measuring equipment   69
5.1 Vernier instruments   5.2 Micrometers   5.3 Dial indicators

6 Cutting tools and cutting fluids   85
6.1 Cutting-tool materials   6.2 Cutting tools   6.3 Cutting-tool
maintenance   6.4 Cutting speed   6.5 Cutting fluids   6.6 Types of
cutting fluid   6.7 Application of cutting fluids   6.8 Safety in the use
of cutting fluids

7 Drilling   102
7.1 The sensitive drilling machine   7.2 Tool holding
7.3 Clamping   7.4 Cutting tools on drilling machines   7.5 Drilling
operations   7.6 Drilling sheet metal   7.7 Drilling plastics

8 Shaping   113
8.1 The shaping machine   8.2 Controls   8.3 Shaping operations

9 Turning   121
9.1 Centre-lathe elements   9.2 Centre-lathe controls
9.3 Workholding   9.4 Centre-lathe operations   9.5 Taper turning
9.6 Screw-cutting

10 Surface grinding   140
10.1 Elements of a surface-grinding machine   10.2 Controls
10.3 Workholding   10.4 Grinding wheels   10.5 Surface-grinding
operations

11 Milling   153
11.1 Milling-machine elements   11.2 Controls   11.3 Milling
cutters   11.4 Cutter mounting   11.5 Workholding   11.6 Milling
operations

12 Joining methods   169
12.1 Mechanical fasteners   12.2 Screw threads   12.3 Locking
devices   12.4 Riveting   12.5 Soft soldering   12.6 Solders
12.7 Brazing   12.8 Welding   12.9 Adhesives   12.10 Electrical
connections   12.11 Relative merits of joining methods

13 Materials   189
13.1 Physical properties   13.2 Mechanical properties
13.3 Comparison of properties   13.4 Plain-carbon steel   13.5 Heat
treatment of plain-carbon steel   13.6 Cast iron   13.7 Copper and its
alloys   13.8 Aluminium and its alloys   13.9 Die-casting alloys
13.10 Lead   13.11 Contact metals   13.12 Bearing materials
13.13 Metal protection   13.14 Corrosion   13.15 Protective coatings
13.16 Painting

14 Plastics   213
14.1 Thermoplastics and thermosetting plastics   14.2 Types
of plastics   14.3 Working in plastics   14.4 Welding
14.5 Machining   14.6 Heat bending   14.7 Encapsulation
14.8 Plastics moulding processes

15 Primary forming processes   232
15.1 Forms of supply of raw materials   15.2 Properties of raw
materials   15.3 Sand casting   15.4 Rolling   15.5 Extrusion
15.6 Drawing   15.7 Forging   15.8 Selection of a primary process

16 Presswork   243
16.1 Presses   16.2 Press-tool design   16.3 Blanking, piercing and
bending operations   16.4 Blanking layouts

17 Investment casting and shell moulding   259
17.1 Investment casting   17.2 Metals for investment casting
7.3 Shell moulding

18 Die-casting   267
18.1 Gravity die-casting   18.2 Low-pressure die-casting   18.3 High-
pressure die-casting   18.4 Die-casting metals   18.5 Special features
of die-castings   18.6 Advantages and limitations of die-castings
18.7 Choice of a die-casting process

Appendices   281

Index   285

Preface

Preparing the second edition has enabled me to update a number of areas and to increase the scope of the book by including additional material. It has also afforded the opportunity of resetting to current popular book size and format.

In this second edition I have increased the content to cover a wider range of topics in order to make the book even more comprehensive by providing additional chapters on processes to include sand casting, rolling, extrusion, drawing, forging, presswork, investment casting, shell moulding and die casting.

I have updated the Safe Practices chapter to include current Health and Safety Regulations and the chapter on Measuring Equipment to include electronic instruments. A section on bonded abrasive grinding wheels has been added to the chapter on Surface Grinding and moulding processes has been included in the chapter on Plastics.

Acknowledgements

The author and publishers would like to thank the following organisations for their kind permission to reproduce photographs or illustrations:

Chubb Fire Ltd (figs 1.2–4); Desoutter Brothers Ltd (fig. 2.21); Neill Tools Ltd (figs 3.10, 3.11, 3.14, 5.17, 5.13); Mitutoyo (UK) Ltd (figs 3.15, 3.20, 5.4, 5.7, 5.8, 5.10, 5.11, 5.12, 5.20, 5.21–4, 5.26, 5.27, 5.29, 5.30, 5.31); A.J. Morgan & Son (Lye) Ltd (figs 4.2, 4.8); Walton and Radcliffe (Sales) Ltd (fig. 4.3); Q-Max (Electronics) Ltd (fig. 4.4); T. Norton & Co. Ltd (figs 4.5, 16.1); Thomas Mercer Ltd (figs 5.28, 5.32); W.J. Meddings (Sales) Ltd (fig. 7.1); Gate Machinery Co. Ltd (fig. 8.1); T.S. Harris & Sons Ltd (fig. 9.1); Pratt Burnerd International Ltd (figs 9.8–13); Elliot Machine Tools Ltd (figs 10.1–2); James Neill (Sheffield) Ltd (figs 10.8, 10.9); Clarkson International Tools Ltd (fig. 11.10); Hinchley Engineering Co. Ltd (fig. 14.7); Dow Corning Ltd (fig. 14.8); Sweeny and Blockside (Power Pressing) (fig. 16.2); Verson International Ltd (fig. 16.3), P.J. Hare Ltd (fig. 16.4); Lloyd Colley Ltd (fig. 16.13); P I Castings (Altringham) (figs 17.1–7); Dennis Castings (fig. 17.9); and Lloyds British Testing Ltd for information on lifting equipment.

1 Safe practices

Almost everyone working in a factory has at some stage in his or her career suffered an injury requiring some kind of treatment or first aid. It may have been a cut finger or something more serious. The cause may have been carelessness by the victim or a colleague, defective safety equipment, not using the safety equipment provided, or inadequate protective clothing. Whatever the explanation given for the accident, the true cause was most likely a failure to think ahead. You must learn to work safely. Your workplace will have its own safety rules so obey them at all times. Ask if you don't understand any instruction and do report anything which seems dangerous, damaged or faulty.

1.1 Health and Safety at Work Act 1974 (HSW Act)

This Act of Parliament came into force in April 1975 and covers all people at work except domestic servants in a private household. It is aimed at people and their activities, rather than at factories and the processes carried out within them.

The purpose of the Act is to provide a legal framework to encourage high standards of health and safety at work.

Its aims are:

- to secure the health, safety, and welfare of people at work;
- to protect other people against risks to health or safety arising from the activity of people at work;
- to control the keeping and use of dangerous substances and prevent people from unlawfully having or using them;
- control the release into the atmosphere of noxious or offensive substances, from prescribed premises.

1.2 Health and safety organisation (Fig. 1.1)

The HSW Act established two bodies, the Health and Safety Commission and the Health and Safety Executive.

Most of the health and safety regulations are the responsibility of the Secretary of State for Employment. These regulations are normally based on proposals submitted by the Health and Safety Commission after consultation with organisations representing, among others, employees, employers, local authorities, and professional bodies.

The Health and Safety Commission consists of representatives from both sides of industry, and from local authorities, and is responsible for developing policies in health and safety.

The Health and Safety Executive is appointed by the Commission with the approval of the Secretary of State and is responsible for enforcing legal requirements, as well as providing an advisory service to both sides of industry.

The Executive also appoints inspectors to carry out its enforcement functions.

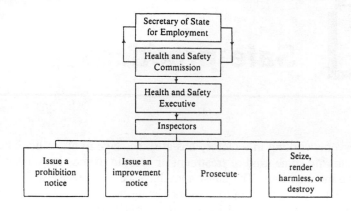

Figure 1.1 Health and safety organisation

Inspectors may visit a workplace without notice. They may want to investigate an accident or complaint, or examine the safety, health and welfare aspects of the business. They have the right to talk to employees and safety representatives and to take photographs and samples.

If there is a problem an inspector can:

- issue a prohibition notice to stop any activity which could result in serious personal injury, until remedial action is taken;
- issue an improvement notice requiring a fault to be remedied within a specified time;
- prosecute any person who does not comply with the regulations – this can lead to a fine, imprisonment, or both;
- seize, render harmless or destroy any substance or article considered to be the cause of imminent danger or serious personal injury.

1.3 Employer's responsibilities (Fig. 1.2)

Employers have a general duty under the HSW Act 'to ensure, so far as is reasonably practicable, the health, safety and welfare at work of their employees'. The HSW Act

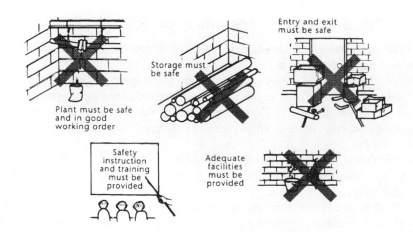

Figure 1.2 Duties of employers

specifies five areas which in particular are covered by the employers general duty.

1. To provide and maintain machinery, equipment and other plant, and systems of work that are safe and without risk to health. ('Systems of work' means the way in which the work is organised and includes layout of the workplace, the order in which jobs are carried out, or special precautions to be taken before carrying out certain hazardous tasks.)
2. Ensure ways in which particular articles and substances (e.g. machinery and chemicals) are used, handled, stored and transported are safe and without risk to health.
3. Provide information, instruction, training and supervision necessary to ensure health and safety at work. *Information* means the background knowledge needed to put the instruction and training into context. *Instruction* is when someone shows others how to do something by practical demonstration. *Training* means having employees practise a task to improve their performance. *Supervision* is needed to oversee and guide in all matters related to the task.
4. Ensure any place under their control and where their employees work is kept in a safe condition and does not pose a risk to health. This includes ways into and out of the workplace.
5. Ensure the health and safety of their employees' working environment (e.g. heating, lighting, ventilation, etc.). They must also provide adequate arrangements for the welfare at work of their employees (the term 'welfare at work' covers facilities such as seating, washing, toilets, etc.).

1.4 Safety policy

The HSW Act requires every employer employing more than five people to prepare a written statement of their safety policy. The written policy statement must set out the employers' aims and objectives for improving health and safety at work.

The purpose of a safety policy is to ensure that employers think carefully about hazards at the workplace and about what should be done to reduce those hazards to make the workplace safe and healthy for their employees.

Another purpose is to make employees aware of what policies and arrangements are being made for their safety. For this reason you must be given a copy which you must read, understand and follow.

The written policy statement needs to be reviewed and revised jointly by employer and employees' representatives as appropriate working conditions change or new hazards arise.

1.5 Safety Representatives and Safety Committees Regulations 1977

Safety representatives
The Regulations came into force on 1 October 1978 and provide recognised trade unions with the right to appoint safety representatives to represent the employees in consultations with their employers about health and safety matters of the organisation.

The HSW Act requires every employer to consult safety representatives in order to make and maintain arrangements to enable the employer and the employees to cooperate in the promotion and development of health and safety measures and to check their effectiveness.

An employer must give safety representatives the necessary time off, with pay, to carry out their functions and receive appropriate training.

The functions of a safety representative include:

- investigating potential hazards and dangerous occurrences in the workplace;
- investigating complaints relating to an employee's health, safety or welfare at work;
- making representations to the employer on matters affecting the health, safety or welfare of employees at the workplace;
- carrying out inspections of the workplace where there has been a change in conditions of work, or there has been a notifiable accident or dangerous occurrence in a workplace or a notifiable disease has been contracted there;
- representing the employees he or she was appointed to represent in consultation with inspectors or any enforcing authority;
- attending meetings of safety committees.

Safety committees

The HSW Act requires an employer to establish a safety committee if requested in writing by at least two safety representatives. The main objective of such a committee is to promote cooperation between employers and employees in setting up, developing and carrying out measures to ensure the health and safety at work of the employees. Its functions can include:

- studying safety and accident reports so that unsafe and unhealthy conditions and practices may be identified and recommendations made for corrective action;
- considering reports by inspectors and by safety representatives;
- assisting in developing works safety rules and safe systems of work;
- monitoring the effectiveness of employee safety training;
- monitoring the adequacy of health and safety communication and publicity in the workplace;
- providing a link with the appropriate enforcing agency.

1.6 Employees' responsibilities (Fig. 1.3)

Under the HSW Act it is the duty of every employee while at work:

- To take reasonable care for their own health and safety and that of others who may be affected by what they do or don't do.

  This duty implies not only avoiding silly or reckless behaviour but also understanding hazards and complying with safety rules and procedures. This means that you correctly use all work items provided by your employer in accordance with the training and instruction you received to enable you to use them safely.
- To cooperate with their employer on health and safety.

  This duty means that your should inform, without delay, of any work situation which might be dangerous and notify any shortcomings in health and safety arrangements so that remedial action may be taken.

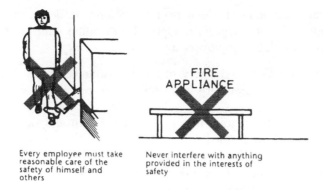

Every employee must take reasonable care of the safety of himself and others

Never interfere with anything provided in the interests of safety

Figure 1.3 Duties of employees

The HSW Act also imposes a duty on all people, both people at work and members of the public, including children to not intentionally interfere with or misuse anything that has been provided in the interests of health, safety and welfare.

The type of things covered include fire escapes and fire extinguishers, perimeter fencing, warning notices, protective clothing, guards on machinery and special containers for dangerous substances.

You can seen that it is essential for you to adopt a positive attitude and approach to health and safety in order to avoid, prevent and reduce risks at work. Your training is an important way of achieving this and contributes not only to your own, but to the whole organisation's, health and safety culture.

1.7 New regulations for health and safety at work

Six new sets of health and safety at work regulations came into force on 1 January 1993. The new regulations implement European Community (EC) directives on health and safety at work in the move towards a single European Union. At the same time they are part of a continuing modernisation of existing UK law.

Most of the duties in the new regulations are not completely new but clarify and make more explicit what is in current health and safety law. A lot of out-of-date law will be repealed by the new regulations, for example many parts of the Factories Act 1961.

The six regulations are:

- Management of Health and Safety at Work Regulations 1992;
- Provision and Use of Work Equipment Regulations 1992;
- Workplace (Health, Safety and Welfare) Regulations 1992;
- Personal Protective Equipment at Work Regulations 1992;
- Health and Safety (Display Screen Equipment) Regulations 1992 (covers computer monitors and is not relevant to this book);
- Manual Handling Operations Regulations 1992

1.8 Management of Health and Safety at Work Regulations 1992

These Regulations set out broad general duties which operate with the more specific ones in other health and safety regulations. They are aimed mainly at improving health and safety management. Their main provisions are designed to encourage a more systematic and better organised approach to dealing with health and safety.

The Regulations require employers to:

- assess the risk to health and safety of employees and anyone else who may be affected so that the necessary preventive and protective measures can be identified;
- make arrangements for putting into practice the health and safety measures that follow from the risk assessment;
- provide appropriate health surveillance of employees where necessary;
- appoint competent people to help devise and apply the measures needed;
- set up emergency procedures;
- give employees information about health and safety matters;
- make sure that employees have adequate health and safety training and are capable enough at their jobs to avoid risk;
- co-operate with any other employers who share a work site;
- give some health and safety information to temporary workers, to meet their special needs.

The Regulations also:

- place a duty on employees to follow health and safety instructions and report danger, and;
- extend the current law which requires employers to consult employees safety representatives and provide facilities for them.

1.9 Provision and Use of Work Equipment Regulations 1992 (PUWER)

These Regulations lay down important health and safety laws for the provision and use of work equipment and are designed to pull together and tidy up the laws governing equipment used at work. Much old legislation including seven sections of the Factories Act 1961 has been replaced. Its primary objective is to ensure the provision of safe work equipment and its safe use.

These Regulations came into force on 1 January 1993 and will operate alongside the HSW Act. Some of the Regulations did not apply to certain categories of work equipment until 1 January 1997.

Work equipment has wide meaning and is broadly defined to include anything from a hand tool, through machines of all kinds, to a complete plant such as a refinery.

PUWER cover the health and safety requirement in respect of the following.

- *The suitability of work equipment* – equipment must be suitable by design and construction for the actual work it is provided to do.
- *Maintenance of work equipment in good repair* – from simple checks on hand tools such as loose hammer heads to specific checks on lifts and hoists. When maintenance work is carried out it should be done in safety and without risk to health.
- *Information and instruction on use of the work equipment* – including instruction sheets, manuals or warning labels from manufacturers or suppliers. Adequate

Figure 1.4 Guard fitted to horizontal milling machine

training for the purposes of health and safety in the use of specific work equipment.
- *Dangerous parts of machinery* – guarding machinery to avoid the risks arising from mechanical hazards. The principal duty is to take effective measures to prevent contact with dangerous parts of machinery by providing:
  - i) fixed enclosing guards;
  - ii) other guards (see Fig. 1.4) or protection devices;
  - iii) protection appliances (jigs, holders);
  - iv) information, instruction, training and supervision.
- *Protection against specified hazards*
  - i) material falling from equipment;
  - ii) material ejected from a machine;
  - iii) parts of the equipment breaking off e.g. grinding wheel bursting;
  - iv) parts of equipment collapsing e.g. scaffolding;
  - v) overheating or fire e.g. bearing running hot, ignition by welding torch;
  - vi) explosion of equipment e.g. failure of a pressure-relief device;
  - vii) explosion of substance in the equipment e.g. ignition of dust.
- *High and very low temperature* – prevent the risk of injury from contact with hot (blast furnace, steam pipes) or very cold work equipment (cold store).
- *Controls and control systems* – starting work equipment should only be possible by using a control and it should not be possible for it to be accidentally or inadvertently operated nor 'operate itself' (by vibration or failure of a spring mechanism).

Stop controls should bring the equipment to a safe condition in a safe manner. Emergency stop controls are intended to effect a rapid response to potentially dangerous situations and should be easily reached and activated. Common types are mushroom headed buttons (see Fig. 1.5), bars, levers, kick plates or pressure-sensitive cables.

It should be possible to identify easily what each control does. Both the controls and their markings should be clearly visible and factors such as colour, shape and position are important.

Figure 1.5 Mushroom-headed stop button

- *Isolation from source of energy* – to allow equipment to be made safe under particular circumstances, for example when maintenance is to be carried out or when an unsafe condition develops. Isolation may be achieved by simply removing a plug from an electrical socket or by operating an isolating switch or valve.

  Sources of energy may be electrical, pressure (hydraulic or pneumatic) or heat.
- *Stability* – there are many types of work equipment that might fall over, collapse or overturn unless they are fixed. Most machines used in a fixed position should be bolted down. Some types or work equipment such as mobile cranes may need counterbalance weights.

  Ladders should be at the correct angle (a slope of four units up to each one out from the base), correct height (at least 1 metre above the landing place) and tied at the top or secured at the foot.
- *Lighting* – if the lighting in the workplace is insufficient for detailed tasks then additional lighting will need to be provided, for example local lighting on a machine (Fig. 1.6).
- *Markings* – there are many instances where marking of equipment is appropriate for health and safety reasons, for example start/stop controls, safe working load on cranes or types of fire extinguishers.

Figure 1.6 Local lighting on a centre lathe

- *Warnings* – normally in the form of a permanent printed notice or similar, for example: 'head protection must be worn' (see page 20). Portable warnings are also necessary during temporary operations such as maintenance.

  Warning devices can be used which may be audible, for example reversing alarms on heavy vehicles, or visible, for example lights on a control panel. They may indicate imminent danger, development of a fault or the continued presence of a potential hazard.

  They must all be easy to see and understand, and they must be unambiguous.

1.10 Workplace (Health, Safety and Welfare) Regulations 1992

These Regulations will also tidy up a lot of existing requirements. They will replace many pieces of old law, including parts of the Factories Act 1961. They will be much easier to understand making it clearer what is expected of everyone. They came into force on 1 January 1993 but for existing workplaces the Regulations took effect on 1 January 1996.

These Regulations set general requirements which are listed here in four broad areas:

- Working environment
  - i) ventilation
  - ii) temperature in indoor workplace
  - iii) lighting including emergency lighting
  - iv) room dimensions and space
  - v) suitability of workstations and seating.
- *Safety*
  - i) safe passage of pedestrians and vehicles (e.g. traffic routes, must be wide enough and marked where necessary, and there must be enough of them).
  - ii) windows and skylights (safe opening, closing and cleaning).
  - iii) transparent or translucent surfaces in doors and partitions (use of safety material and marking)
  - iv) doors, gates and escalators (safety devices)
  - v) floors (construction and maintenance, obstructions and slipping and tripping hazards)
  - vi) falling from heights and into dangerous substances
  - vii) falling objects.
- *Facilities*
  - i) toilets
  - ii) washing, eating and changing facilities
  - iii) clothing storage
  - iv) drinking water
  - v) rest areas (and arrangements to protect people from the discomfort of tobacco smoke).
- *Housekeeping*
  - i) maintenance of workplace, equipment and facilities
  - ii) cleanliness
  - iii) removal of waste materials.

1.11 Personal Protective Equipment at Work Regulations 1992

These Regulations came into force on 1 January 1993 and set out in legislation, sound principles of selecting, providing and using personal protective equipment (PPE). They replace parts of over 20 old pieces of law (e.g. The Protection of Eyes Regulations 1974 has been revoked). They do not replace the recently introduced laws dealing with PPE (e.g. Control of Substances Hazardous to Health or Noise at Work Regulations).

PPE should always be relied upon as a last resort to protect against risks to health and safety. Engineering controls and safe systems of work should always be considered first. Where the risks are not adequately controlled by other means, the employer has a duty to ensure that suitable PPE is provided, free of charge. PPE will only be suitable if it is appropriate for the risks and the working conditions; takes account of the workers' needs and fits properly; gives adequate protection; and is compatible with any other item of PPE worn.

The employer also has duties to:

- assess the risks and PPE intended to be issued and that it is suitable;
- maintain, clean and replace PPE;
- provide storage for PPE when it is not being used;
- ensure that PPE is properly used; and
- give training, information and instruction to employees on the use of PPE and how to look after it.

PPE is defined as all equipment which is intended to be worn or held to protect against risk to health and safety. This includes most types of protective clothing and equipment such as: eye, head, foot and hand protection; and protective clothing for the body. It does not include ear protectors and respirators which are covered by separate existing regulations.

Eye protection: Serves as a guard against the hazards of impact, splashes from chemicals or molten metal, liquid droplets (chemical mists and sprays), dust, gases and welding arcs. Eye protectors include safety spectacles, eye-shields, goggles, welding filters, face shields and hoods (Fig 1.7).

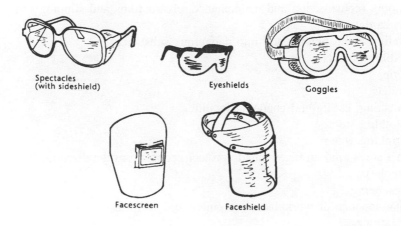

Spectacles (with sideshield)   Eyeshields   Goggles

Facescreen   Faceshield

Figure 1.7 Eye protection

Head protection: Includes industrial safety helmets to protect against falling objects or impact with fixed objects; industrial scalp protectors to protect against striking fixed obstacles, scalping or entanglement; and caps and hairnets to protect against scalping and entanglement.

Foot protection: Includes safety boots or shoes with steel toe caps; foundry boots with steel toe caps, which are heat resistance and designed to keep out molten metal; wellington boots to protect against water and wet conditions; and anti-static footwear to prevent the build up of static electricity on the wearer.

Hand protection: Gloves of various design provide protection against a range of hazards including cuts and abrasions; extremes of temperature (hot and cold); skin irritation and dermatitis; and contact with toxic or corrosive liquids. Barrier creams may sometimes be used as an aid to skin hygiene in situations where gloves cannot be used.

Protective clothing: Types of clothing used for body protection include coveralls, overalls and aprons to protect against chemicals and other hazardous substances; outfits to protect against cold, heat and bad weather; and clothing to protect against machinery such as chain saws. Types of clothing worn on the body to protect the person include high visibility clothing; life-jackets and buoyancy aids.

1.12 The Manual Handling Operations Regulations 1992

These Regulations came into force on 1 January 1993. The Regulations apply to the manual handling of loads and seek to prevent injury, not only to the back, but to any part of the body. Account is taken of physical properties of loads which may affect grip or cause injury by slipping, roughness, sharp edges or extremes of temperature.

The regulations require that where there is the possibility of risk to employees from the manual handling of loads, the employer should take the following measures, in this order:

1. avoid hazardous manual handling operations so far as is reasonably practical;
2. assess any hazardous manual handling operations that cannot be avoided; and
3. reduce the risk of injury so far as is reasonably practicable.

Steps taken to avoid manual handling or reduce the risk of injury must be regularly checked to see if they are effective.

It is a requirement of the HSW Act and the Management of Health and Safety at Work Regulations 1992 that employers provide their employees with health and safety information and training. This should include specific information and training on manual handling, injury risk and prevention, as part of the steps to reduce risks required by these Regulations.

Although the Regulations do not set out specific requirements such as weight limits they do give numerical guidelines to assist with assessment. Guidelines for lifting and lowering are shown in Fig. 1.8. This shows guideline figures taking into consideration vertical and horizontal position of the hands as they move during the handling operation, e.g. 10 kg if lifted to shoulder height at arm's length or 5 kg if lifted to full height at arm's length. This assumes that the load can be easily grasped, with a good body position and in reasonable working conditions. If the hands enter

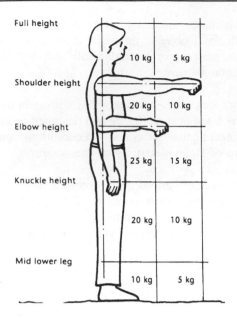

Figure 1.8 Lifting and lowering loads

more than one of the boxes during lifting, the smallest weight figure should be used.

1.13 Good handling techniques

The development of good handling technique is no substitute for the risk reduction steps already outlined but is an important addition which requires training and practice. The following should form the basic lifting operation.

Stop and think: Plan the lift. Organise the work to minimise the amount of lifting necessary. Know where you are going to place the load. Use mechanical assistance if possible. Get help if load is too heavy. Make sure your path is clear. Don't let the load obstruct your view. For a long lift, i.e. from floor to shoulder height, consider a rest mid-way on a bench in order to adjust your grip. Alternatively lift from floor to knee then from knee to carrying position – reverse this method when setting the load down.

Place your feet: Keep your feet apart to give a balanced and stable base for lifting (see Fig. 1.9). Your leading leg should be as far forward as is comfortable.

Figure 1.9 Placing the feet

Figure 1.10 Adopting a good posture

Figure 1.11 Keep load close to body

Adopt a good posture: Keep your back straight and upright. Bend your knees and let your legs do the work (see Fig. 1.10). Keep your shoulders level and facing the same direction as your hips, i.e. don't twist your body.

Get a firm grip: Lean forward a little over the load to get a good grip. Try to keep your arms within the boundary formed by the legs. Balance the load using both hands. A hook grip is less tiring than keeping the fingers straight. Wear gloves if the surface is rough or has sharp edges. Take great care if load is wrapped or slippery in any way.

Don't jerk: Carry out the lifting operation smoothly, keeping control of the load. Move the feet: don't twist your body if you turn to the side.

Keep close to the load: If the load is not close when lifting, try sliding it towards you. Keep the load close to your body for as long as possible. Keep the heaviest side of the load next to your body (see Fig. 1.11).

Put down: Putting down the load is the exact reverse of your lifting procedure. If precise positioning of the load is required, put it down, then slide it to the desired position.

   Whenever possible make use of mechanical assistance involving the use of handling aids. Although an element of manual handling is still present, body forces are applied more efficiently reducing the risk of injury. Levers can be used which lessen the body force required and also remove fingers from a potentially dangerous area. Hand- or power-operated hoists can be used to support a load and leave the operator free to control its positioning. A trolley, sack truck or roller conveyer can reduce the effort required to move a load horizontally while chutes using gravity can be used from one height to the next. Hand-held hooks and suction pads can be used where loads are difficult to grasp. (As a general rule, loads over 20 kg need the assistance of lifting gear.)

1.14 The Reporting of Injuries, Diseases and Dangerous Occurrences Regulations 1985 (RIDDOR)

RIDDOR came into effect in April 1986. These Regulations require injuries, diseases and occurrences in specified categories to be notified to the relevant enforcing authority. In the case of a factory, the enforcing authority is the Health and Safety Executive.

The enforcing authority must be notified without delay, normally by a phone call, followed by a written report within seven days.

Immediate notification is required for the following:

- any fatal injuries to employees or other people in an accident connected with your business;
- any major injuries to employees or other people in an accident connected with your business (major injuries include fractures, amputation, loss of sight, injury from electric shock and any other injury which results in the person being admitted to hospital for more than 24 hours). A written report must be sent within seven days of any other injury to an employee which results in an absence of more than three working days;
- any of the dangerous occurrences listed in the Regulations (these include the collapse, overturning or failure of lifts, hoists and cranes, explosion of vessels, electrical fires, the sudden release of highly flammable liquids);
- report notifying specific disease related to particular work activities listed in the Regulations (the general diseases covered include certain poisonings, some skin diseases, lung diseases, infections and other conditions such as occupational cancer).

A record must be kept of any injury, occurrence or case of disease requiring report. This should include the date, time and place, personal details of those involved and a brief description of the nature of the event.

1.15 Noise at Work Regulations 1989

These Regulations are intended to reduce hearing damage caused by loud noise. Exposure levels can cause incurable hearing damage. The important factors are: (1) the noise level, given the decibel units dB(A) and (2) how long the person is exposed to the noise; daily, or over a number of years.

Action levels are set and action has to be taken when they are reached:

- first action level 85 dB(A);
- second action level 90 dB(A) (these are personal daily exposure levels and are denoted $L_{EP,d}$);
- peak action level equivalent to 140 dB(A) (where cartridge-operated tools are used even occasionally).

Example levels are shown in Table 1.1.

The need to wear ear protection should be the last resort. The best protection against noise is to control it at source by designing or choosing machines and processes to make less noise, by enclosing noisy machines or putting in a separate room, or by fitting silencers. Finally, if all else fails, ear protection should be provided in noisy areas.

The Regulations require an employer to:

Table 1.1

Noise source	Noise level dB(A)
Domestic food blender	81
Electric drill	87
Sheet metal shop	93
Circular saw	99
Chain saw	102
Hand grinding metal	108
Jet aircraft taking off 25 metres away	140

- assess noise levels;
- inform workers at the first action level about the risks to hearing and provide ear protectors;
- control noise exposure if noise reaches second or peak action levels;
- mark ear protection zones with notices and make sure that everyone wears ear protectors.

1.16 Electrical hazards

Electrical equipment of some kind is used in every factory. Electricity should be treated with respect – it cannot be seen or heard, but it can kill. Even if it is not fatal, serious disablement can result through shock and burns. Also, a great deal of damage to property and goods can be caused, usually through fire or explosion as a result of faulty wiring or faulty equipment.

The Electricity at Work Regulations 1989 came into force on 1 April 1990. The purpose of the Regulation is to require precautions to be taken against the risk of death or personal injury from electricity in work activities.

The Institution of Electrical Engineers Regulations for electrical installations (IEE Wiring Regulations), although non-statutory, is widely recognised and accepted in the UK and compliance with these is likely to mean compliance with the relevant parts of the Electricity at Work Regulations 1989.

The major hazards arising from the use of electrical equipment are:

Electric shock: The body responds in a number of ways to electric current flowing through it, any one of which can be fatal. The chance of electric shock is increased in wet or damp conditions, or close to conductors such as working in a metal tank. Hot environments where sweat or humidity reduce the insulation protection offered by clothing increase the risk.

Electric burn: This is due to the heating effect caused by electric current passing through body tissue, most often the skin at the point of contact giving rise to the electric shock.

Fire: Caused by electricity in a number of ways including: overheating of cables and electrical equipment due to overloading; leakage currents due to poor or inadequate insulation; overheating of flammable materials placed too close to electrical equipment; ignition of flammable materials by sparking of electrical equipment.

Arcing: Generates ultra-violet radiation causing a particular type of burn similar to severe sunburn. Molten metal resulting from arcing can penetrate, burn and lodge in the flesh. Ultra-violet radiation can also cause damage to sensitive skin and to eyes, e.g. arc eye in metal arc welding.

Explosion: These include the explosion of electrical equipment, e.g. switchgear or motors, or where electricity causes the ignition of flammable vapours, gases, liquids and dust by electric sparks or high temperature electrical equipment.

Electrical precautions

Where it is possible for electrical equipment to become dangerous if a fault should arise, then precautions must be taken to prevent injury. These precautions include:

Double insulation: The principle is that the live conductors are covered by two discrete layers of insulation. Each layer would provide adequate insulation in itself but together they ensure little likelihood of danger arising from insulation failure. This arrangement avoids the need for an earth wire. Double insulation is particularly suitable for portable equipment such as drills. However, safety depends on the insulation remaining in sound condition and the equipment must be properly constructed, used and maintained.

Earthing: In the UK the electricity supply is connected to earth. It is this system that enables earth faults on electrical equipment to be detected and the electrical supply to be cut off automatically. This automatic cut-off is performed by fuses or automatic circuit breakers: if a fault occurs the fuse will blow and break the circuit. Although they do not eliminate the risk of electric shock, danger may be reduced by the use of a residual current device (RCD) designed to operate rapidly at small leakage currents. RCDs should only be considered as providing a second line of defence. It is essential to regularly operate the test trip button to maintain their effectiveness.

Use of safe voltages: Reduced voltage systems (110 volts) are particularly suitable for portable electrical equipment in construction work and in high conducting locations such as boilers, tunnels and tanks; where the risk to equipment and trailing cables is high; and where the body may be damp.

The human body as part of a circuit

In order to minimise the risk of shock and fire, any metalwork other than the current-carrying conductor must be connected to earth. The neutral of the electrical supply is earthed at the source of distribution, i.e. the supply transformer, so that, if all appliances are also connected to earth, a return path for the current will be available through earth when a fault occurs (see Fig. 1.12). To be effective, this earth path must be of sufficiently low resistance to pass a relatively high current when a fault occurs. This higher current will in turn operate the safety device in the circuit, i.e. the fuse will blow.

Accidents happen when the body provides a direct connection between the live conductors – when the body or a tool touches equipment connected to the supply. More often, however, the connection of the human body is between one live conductor and earth, through the floor or adjacent metalwork (see Fig. 1.13). Metal

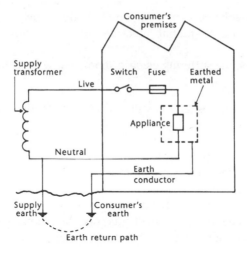

Figure 1.12 Electric circuit for premises

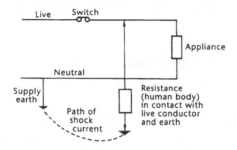

Figure 1.13 Human body as a resistance in electric circuit

pipes carrying water, gas, or steam, concrete floors, radiators, and machine structures all readily provide a conducting path of this kind.

Any article of clothing containing any metal parts increases the likelihood of accidental electrical contact. Metal fittings such as buttons, buckles, metal watch or wrist bracelets or dog tags, and even rings could result in shock or burns.

Wetness or moisture at surfaces increases the possibility of leakage of electricity, by lowering the resistance and thus increasing the current. Contact under these conditions therefore increases the risk of electric shock.

All metals are good electrical conductors and therefore all metallic tools are conductors. Any tool brought near a current carrying conductor will bring about the possibility of a shock. Even tools with insulated handles do not guarantee that the user will not suffer shock or burns.

Electric shock and treatment

If the human body accidentally comes in contact with an electrical conductor which is connected to the supply, a current may, depending on the conditions, flow through the body. This current will at least produce violent muscular spasms which may cause the body to be flung across the room or fall off a ladder. In extreme cases the heart will stop beating.

Burns are caused by the current acting on the body tissue and internal heating can also take place leading to partial blockage of blood flow through the blood vessels.

In the event of someone suffering electric shock, know what to do – it should form part of your training.

1. Shout for help – if the casualty is still in contact with electric current, switch off or remove the plug.
2. If the current cannot be switched off, take special care to stand on a dry non-conducting surface and pull or push the victim clear using a length of dry cloth, jacket, or a broom. Remember: do not touch the casualty as you will complete the circuit and also receive a shock.
3. Once free, if the casualty is breathing, put in recovery position and get the casualty to hospital; if the casualty is not breathing give mouth-to-mouth resuscitation, check pulse, and, if absent, apply chest compressions and call for medical assistance.

Posters giving the detailed procedure to be followed in the event of a person suffering electric shock must be permanently displayed in your workplace. With this and your training you should be fully conversant with the procedures – remember it could save a life.

General electrical safety rules

- Ensure that a properly wired plug is used for all portable electrical equipment (see Fig. 1.14)

brown wire	**live** conductor
blue wire	**neutral** conductor
green/yellow wire	**earth** conductor.

- Never improvise by jamming wires in sockets with nails or matches.
- Moulded rubber plugs are preferable to the brittle plastic types, since they are less prone to damage.
- All electrical connections must be secure, loose wires or connections can arc.
- A fuse of the correct rating must be fitted – this is your safeguard if a fault develops – never use makeshift fuses such as pieces of wire.
- Any external metal parts must be earthed so that if a fault develops, the fuse will blow and interrupt the supply.
- Never run power tools from lamp sockets.

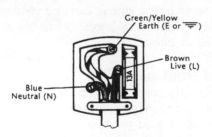

Figure 1.14 Correctly wired plug

- Connection between the plug and equipment should be made with the correct cable suited to the current rating of the equipment.
- Old or damaged cable should never be used.
- Equipment should always be disconnected from the mains supply before making any adjustment, even when changing a lamp.
- Do not, under any circumstances, interfere with any electrical equipment or attempt to repair it yourself. All electrical work should be done by a qualified electrician. A little knowledge is often sufficient to make electrical equipment function but a much higher level of knowledge and expertise is usually needed to ensure safety.

1.17 Safety signs and colours

Colours play an essential safety role in giving information for use in the prevention of accidents, for warning of health hazards, to identify contents of gas cylinders, pipeline and services, the identification and safe use of cables and components in electronic and electrical installations as well as the correct use of fire-fighting equipment.

The purpose of a system of safety colours and safety signs is to draw attention to objects and situations which affect or could affect health and safety. The use of a system of safety colours and safety signs does not replace the need for appropriate accident prevention measures.

British Standard BS5378:Part 1:1980 Safety Signs and Colours is concerned with a system for giving safety information which does not, in general, require the use of words. BS5499:Part 1:1990 extends the basic framework concerning safety colours and safety signs in BS5378 with regard to fire.

Table 1.2

Safety colour	Meaning	Examples of use
Red (white background colour with black symbols)	Stop Prohibition (**Don't** do)	Stop signs Emergency stops Prohibition signs
Red (white symbols and text)	Fire equipment	Position of fire equipment, alarms, hoses, extinguishers, etc.
Yellow (black symbols and text)	Warning (risk of danger)	Indication of hazards (electrical, explosive, radiation, chemical, vehicle, etc.) Warning of threshold, low passages, obstacles
Green (white symbols and text)	Safe condition (the safe way)	Escape routes Emergency exits Emergency showers First-aid and rescue stations
Blue (white symbols and text)	Mandatory action (**MUST** do)	Obligation to wear personal safety equipment

Figure 1.15 Prohibition – indicates certain behaviour is prohibited

Figure 1.16 Fire equipment

Figure 1.17 Warning – indicates warning of possible hazard

Figure 1.18 Safe condition – conveys information about safe conditions

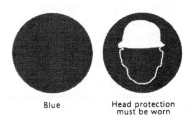

Figure 1.19 Mandatory – indicates specific course of action to be taken

The safety colours, their meaning, and examples of their use are shown in Table 1.2. Examples of the shape and colour of the signs are shown in Figs 1.15–1.19.

Portable fire extinguishers

In most cases the entire body of a portable fire extinguisher is colour coded to indicate the medium contained. Colour coding by medium is intended to provide a means of rapid recognition of the type of extinguisher by trained persons at the time when the extinguisher is needed for use. The use of portable fire extinguishers is dealt with later in this chapter. Table 1.3 shows the extinguishing medium and the corresponding colour code.

Gas containers

Many thousands of people in industry use gas from cylinders in which the gas is contained at high pressure. All users should know and understand the properties of the gas they are using and the correct operating procedures for the equipment being used with the gas. Gas data and safety sheets are readily available from the supplier.

Table 1.3 Colour coding by medium

Extinguishing medium	Colour
Water	Signal red
Foam	Pale cream
Powder	French blue
Carbon dixoide (CO_2)	Black

As well as the container being marked with the gas it contains, colour is also used.

British Standard BS349:1973 'Identification of the Contents of Industrial Gas Containers' sets out the colours used on cylinders to identify the gas contained and therefore relates to safety requirements. There is a range of gases available some of which may be either toxic, flammable or corrosive and the required safety precautions when transporting, handling, storing or using these must be observed.

Table 1.4 shows a range of industrial gases and the corresponding cylinder colour.

Table 1.4

Industrial gas	Cylinder colour
Acetylene	Maroon
Air	French grey
Argon	Peacock blue
Carbon dioxide (CO_2)	Black
Hydrogen	Signal red
Nitrogen	French grey with black band
Oxygen	Black
Propane	Signal red

1.18 Fire

Fire is a phenomenon in which combustible materials, especially organic materials containing carbon, react chemically with the oxygen in the air to produce heat. Flame arises from the combustion of volatile liquids and gases evolved and spreads the fire.

No-one should underestimate the danger of fire. Many materials burn rapidly and the fumes and smoke produced, particularly from synthetic material, including plastics, may be deadly.

There are a number of reasons for fires starting:

- malicious ignition: i.e. deliberate fire raising;
- misuse or faulty electrical equipment: e.g. incorrect plugs and wiring, damaged cables, overloaded sockets and cables, sparking, and equipment such as soldering irons left on and unattended;
- cigarettes and matches: smoking in unauthorised areas, throwing away lighted cigarettes or matches;
- mechanical heat and sparks: e.g. faulty motors, overheated bearings, sparks produced by grinding and cutting operations;

- heating plant: flammable liquids/substances in contact with hot surfaces;
- rubbish burning: casual burning of waste and rubbish.

There are a number of reasons for the spread of fire including:

- delayed discovery;
- presence of large quantities of combustible materials;
- lack of fire separating walls between production and storage areas;
- openings in floors and walls between departments;
- rapid burning of dust deposits;
- oils and fats flowing when burning;
- combustible construction of buildings;
- combustible linings of roofs, ceilings and walls.

Fire prevention

The best prevention is to stop a fire starting:

- where possible use materials which are less flammable;
- minimise the quantities of flammable materials kept in the workplace or store;
- store flammable material safely, well away from hazardous processes or materials, and where appropriate, from buildings;
- warn people of the fire risk by a conspicuous sign at each workplace, storage area and on each container;
- some items, like oil-soaked rags, may ignite spontaneously. Keep them in a metal container away from other flammable material;
- before welding or similar work remove or insulate flammable material and have fire extinguishers to hand;
- control ignition sources, e.g. naked flames and sparks, and make sure that 'no smoking' rules are obeyed;
- do not leave goods or waste to obstruct gangways, exits, stairs, escape routes and fire points;
- make sure that vandals do not have access to flammable waste materials;
- comply with the specific precautions for highly flammable gas cylinders such as acetylene;
- after each spell of work, check the area for smouldering matter or fire;
- burn rubbish in a suitable container well away from buildings and have fire extinguishers to hand;
- never wedge open fire-resistant doors designed to stop the spread of fire and smoke;
- have enough fire extinguishers, of the right type and properly maintained, to deal promptly with small outbreaks of fire.

Fire Precautions Act 1971

Under this Act the owner of a factory employing more than 20 people (more than ten if on a floor above ground level) is required to have a fire certificate from the fire authority which specifies:

- the use of the premises;
- the means of escape in cases of fire;
- the fire-fighting equipment;
- the fire-warning arrangements;

and may also include requirements for:

- maintaining the means of escape and ensuring that they are not obstructed;
- ensuring that employees receive instructions and training in what to do in the case of fire and keeping records of this training;
- limiting the number of persons who may be in the premises at any one time.

Training has a most important bearing on the safety of the occupants of premises in the event of a fire and it may also contribute to reducing the extent of the damage. Training in fire prevention may be responsible for preventing a fire from starting.

Every employee of a firm should be trained:

- to prevent fires;
- in the action to take if fire occurs.

To ensure that all employees, after training, are familiar with, and understand the procedure in the event of a fire, repeated practice is desirable. After initial practices to establish the procedure, practice drills should be held at least twice a year.

Two-and-a-half minutes to complete evacuation is a reasonable standard to aim for, but in factories where there is a danger of rapid fire spread or of explosion, evacuation may need to be completed in less than one minute.

To avoid panic and accidents, fire drills should always be announced.

Fire fighting

Every employee should know where the portable fire extinguishers, the hose reels and the controls for extinguishing are located and how to operate extinguishers in their working area. This training must include the use of extinguishers on simulated fires.

It must be stressed that fire fighting should only be attempted if it is safe to do so and that an escape route must always be available.

It is also essential to emphasise the limits of first-aid fire fighting in order to show the need to attempt this safely and the importance of first raising the alarm.

As previously stated, a fire requires fuel, oxygen (air) and heat. This is shown by the 'fire triangle' in Fig. 1.20, where one side stands for fuel, another for heat and the third for air or oxygen. If any one side is removed the fire inside will go out.

The extinguishing of a fire is generally brought about by depriving the burning substances of oxygen and by cooling them to a temperature below which the reaction is not sustained.

By far the most important extinguishing agent, by reason of its availability and general effectiveness, is water. It is more effective than any other common substance

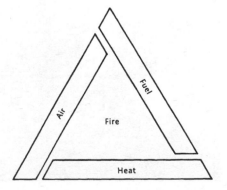

Figure 1.20 Fire triangle

in absorbing heat, thereby reducing the temperature of the burning mass. The steam produced also has a smothering action by lowering the oxygen content of the atmosphere near the fire.

For these reasons the use of a water hose reel in factories is common and is suitable for most fires except those involving flammable liquids or live electrical equipment.

For all practical purposes there are three main classes of fire: A, B and C as well as fires involving electrical equipment and those involving vehicles.

Class A type fires: Fires involving combustible materials such as wood, paper and fabrics.

Class B type fires: Fires involving flammable liquids such as oils, spirits, alcohols, greases, fats and certain plastics.

Class C type fires: Fires involving flammable gases such as propane and butane.

Types of portable fire extinguishers

Water (Fig. 1.21): Colour coded red – these are suitable for class A types of fires. Water is a fast, efficient means of extinguishing these materials and works by having a rapid cooling effect on the fire so that insufficient heat remains to sustain burning and continuous ignition ceases.

Figure 1.21 Water fire extinguisher

Figure 1.22 Spray foam fire extinguisher

Spray foam (Fig. 1.22): Colour coded cream – these are ideal in multi-risk situations where both class A and B type fires are likely. Spray foam has a blanketing effect which both smothers the flame and prevents re-ignition of flammable vapours by sealing the surface of the material. These extinguishers contain an aqueous film-forming foam (AFFF).

Dry powder (Fig. 1.23): Colour coded blue – these are suitable for class A, B and C fires and for vehicle protection. Because dry powder is non-conductive it is ideal for electrical hazards. Dry powder is a highly effective means of extinguishing fires as it interferes with the combustion process and provides rapid fire knockdown.

Figure 1.23 Dry powder fire extinguisher

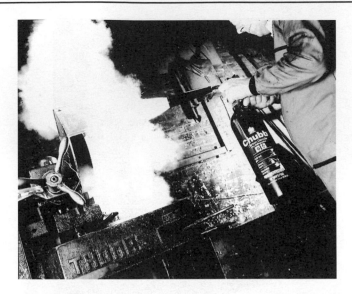

Figure 1.24 Carbon dioxide (CO_2 gas) fire extinguisher

Carbon dioxide (CO_2 gas) (Fig. 1.24): Colour coded black – these are suitable for class B and C type fires. They are also ideal for electrical hazards because CO_2 is non-conductive. CO_2 is an extremely fast fire control medium. These extinguishers deliver a powerful concentration of carbon dioxide gas under great pressure, which smothers the flames very rapidly by displacing air from the local area of the fire. CO_2 is a non-toxic, non-corrosive gas that is harmless to most delicate equipment and materials found in situations such as computer rooms.

Halon: Because of the potential ozone depleting effect, halon portable fire extinguishers are now being withdrawn.

Table 1.5 shows the portable extinguishers most suited to each class of fire.

Table 1.5

Class of fire	Type of extinguisher			
	Water	*Spray foam*	*Dry powder*	*CO_2 gas*
Type A paper, wood, fabric	✔	✔	✔	
Type B flammable liquids		✔	✔	✔
Type C flammable gases			✔	✔
Electrical hazards			✔	✔
Vehicle protection		✔	✔	

1.19 Causes of accidents

Workplace accidents can be prevented – you only need commitment, commonsense and to follow the safety rules set out for your workplace. Safety doesn't just happen – you have to make it happen.

Most accidents are caused by carelessness, through failure to think ahead or as a result of fatigue. Fatigue may be brought on by working long hours without sufficient periods of rest or even through doing a second job in the evening.

Taking medicine can affect peoples' ability to work safely, as can the effects of alcohol. Abuse of drugs or substances such as solvents can also cause accidents at work.

Serious injury and even death have resulted from horseplay, practical jokes or silly tricks. There is no place for this type of behaviour in the workplace.

Improper dress has led to serious injury: wearing trainers instead of safety footwear, and loose cuffs, torn overalls, floppy woollen jumpers, rings, chains, watch straps and long hair to get tangled up.

Don't forget, quite apart from the danger to your own health and safety, you are breaking the law if you fail to wear the appropriate personal protective equipment.

Unguarded or faulty machinery, and tools are other sources of accidents. Again within the health and safety law you must not use such equipment and furthermore it is your duty to report immediately defective equipment.

Accidents can occur as a result of the work place environment, e.g. poor ventilation, temperature too high or too low, bad lighting, unsafe passages, doors, floors, and dangers from falls and falling objects.

They can also occur if the workplace, equipment and facilities are not maintained, are not clean, and rubbish and waste materials are not removed.

Many accidents befall new workers in an organisation, especially the young, and are the result of inexperience, lack of information, instruction, training or supervision all of which it is the duty of the employer to provide.

1.20 General health and safety precautions

As already stated you must adopt a positive attitude and approach to health and safety. Your training is an important way of achieving competence and helps to convert information into healthy and safe working practices.

Remember to observe the following precautions.

- *Horseplay*
  work is not the place for horseplay, practical jokes, or silly tricks.
- *Hygiene*
  i) always wash your hands using suitable hand cleaners and warm water before meals, before and after going to the toilet, and at the end of each shift;
  ii) dry your hands carefully on the clean towels or driers provided – don't wipe them on old rags;
  iii) paraffin, petrol or similar solvents should never be used for skin-cleaning purposes;
  iv) use appropriate barrier cream to protect your skin;
  v) conditioning cream may be needed after washing to replace fatty matter and prevent dryness.
- *Housekeeping*
  i) never throw rubbish on the floor;
  ii) keep gangways and work area free of metal bars, components, etc;
  iii) if oil, or grease is spilled, wipe it up immediately or someone might slip and fall.
- *Moving about*

    i) always walk – never run;
    ii) keep to gangways – never take shortcuts;
    iii) look out for and obey warning notices and safety signs;
    iv) never ride on a vehicle not made to carry passengers, e.g. fork-lift trucks.

- *Personal protective equipment*
    i) use all personal protective clothing and equipment, such as ear and eye protectors, dust masks, overalls, gloves, safety shoes and safety helmets;
    ii) get replacements if damaged or worn.

- *Ladders*
    i) do not use ladders with damaged, missing or loose rungs;
    ii) always position ladders on a firm base and at the correct angle – the height of the top support should be four times the distance out from the base;
    iii) ensure the ladder is long enough – at least one metre above the landing place;
    iv) make sure the ladder is tied at the top or secured at the bottom;
    v) never over-reach from a ladder – be safe, get down and move it;
    vi) take all necessary precautions to avoid vehicles or people hitting the bottom of the ladder.

- *Machinery*
    i) ensure you know how to stop a machine before you set it in motion;
    ii) keep your concentration while the machine is in motion;
    iii) never leave your machine unattended while it is in motion;
    iv) take care not to distract other machine operators;
    v) never clean a machine while it is in motion – always isolate it from the power supply first;
    vi) never clean swarf away with your bare hands – always use a suitable rake;
    vii) keep your hair short or under a cap or hairnet – it can become tangled in drills or rotating shafts;
    viii) avoid loose clothing – wear a snug-fitting boiler suit, done up, and ensure that any neckwear is tucked in and secure;
    ix) do not wear rings, chains, or watches at work – they have caused serious injury when caught accidentally on projections;
    x) do not allow unguarded bar to protrude beyond the end of a machine, e.g. in a centre lathe;
    xi) always ensure that all guards are correctly fitted and in position – remember, guards are fitted on machines to protect you and others from accidentally coming in contact with dangerous moving parts.

- *Harmful substances*
    i) learn to recognise hazard warning signs and labels;
    ii) follow all instructions;
    iii) before you use a substance find out what to do if it spills onto your hand or clothes;
    iv) never eat or drink in the near vicinity;
    v) do not take home any clothes which have become soaked or stained with harmful substances;
    vi) do not put liquids or substances into unlabelled or wrongly labelled bottles or containers.

- *Electricity*
    i) make sure you understand all instructions before using electrical equipment;

ii) do not use electrical equipment for any purpose other than, nor in the area other than the intended one;

iii) always switch off or isolate before connecting or disconnecting any electrical equipment.

- *Compressed air*

   i) only use compressed air if allowed to do so;

   ii) never use compressed air to clean a machine – it may blow in your face or someone else's and cause an injury.

- *Fire*

   i) take care when using flammable substances;

   ii) never smoke in 'no smoking' areas;

   iii) do not throw rubbish, cigarette ends, and matches in corners or under benches;

   iv) always make sure that matches and cigarettes are put out before throwing them away;

   v) know the correct fire drill.

- *First aid*

   i) have first aid treatment for every injury however trivial;

   ii) know the first aid arrangements for your workplace.

2 Hand processes

Hand tools are used to remove small amounts of material, usually from small areas of the workpiece. This may be done because no machine is available, the workpiece is too large to go on a machine, the shape is too intricate, or simply that it would be too expensive to set up a machine to do the work.

Since the use of hand tools is physically tiring, it is important that the amount of material to be removed by hand is kept to an absolute minimum and that the correct tool is chosen for the task. Wherever possible, use should be made of the available powered hand tools, not only to reduce fatigue but also to increase the speed of the operation and so reduce the cost.

2.1 Engineer's files

Files are used to perform a wide variety of tasks, from simple removal of sharp edges to producing intricate shapes where the use of a machine is impracticable. They can be obtained in a variety of shapes and in lengths from 150 mm to 350 mm. When a file has a single series of teeth cut across its face it is known as *single-cut* file, and with two sets of teeth cut across its face it is known as *double-cut* file, Fig. 2.1.

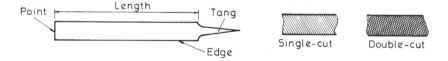

Figure 2.1 Single-cut and double-cut files

The grade of cut of a file refers to the spacing of the teeth and determines the coarseness or smoothness of the file. Three standard grades of cut in common use, from coarsest to smoothest, are *bastard*, *second cut*, and *smooth*. In general, the bastard cut is used for rough filing to remove the most metal in the shortest time, the second cut to bring the work close to finished size, and the smooth cut to give a good finish to the surface while removing the smallest amount of material.

Files are identified either by their general shape – i.e. hand, flat, or pillar – or by their cross-section – i.e. square, three-square, round, half-round, or knife – Fig. 2.2.

Hand file: The hand file is for general use, typically on flat surfaces. It is rectangular in cross-section, parallel in width along its length, but tapers slightly in thickness for approximately the last third of its length towards the point. It is double-cut on both faces, single-cut on one edge, and is plain on the second edge. The plain edge with no teeth is known as the 'safe' edge and is designed to file up to the edge of a surface without damaging it. The taper in thickness enables the file to enter a slot slightly less than its full thickness.

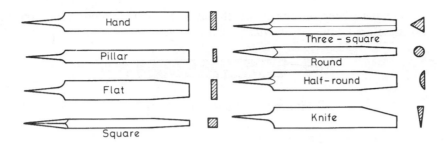

Figure 2.2 Types of file

Pillar file: This file has the same section as a hand file but of a thinner section. It is used for narrow slots and keyways.

Flat file: The flat file is also for general use, typically on flat surfaces. It is rectangular in cross-section and tapers in both width and thickness for approximately the last third of its length towards the point. Both faces are double-cut and both edges single–cut. The tapers in width and thickness enable this file to be used in slots which are narrower than its full width and thickness and which require filing on length and width.

Square file: The square file is of square cross-section, parallel for approximately two thirds of its length, then tapering towards the point. It is double-cut on all sides. This file is used for filing keyways, slots, and the smaller square or rectangular holes with 90° sides.

Three-square file: The three-square or triangular file has a 60° triangle cross-section, parallel for approximately two thirds of its length, then tapering towards the point. The three faces are double-cut and the edges sharp. This file is used for surfaces which meet at less than 90°, angular holes, and recesses.

Round file: The round file is of circular cross-section, parallel for approximately two thirds of its length and then tapering towards the point. Second-cut and smooth files are single-cut, while the bastard is double-cut. This file is used for enlarging round holes, elongating slots, and finishing internal round corners.

Half-round file: The half-round file has one flat and one curved side. It is parallel for approximately two thirds of its length, then tapers in width and thickness towards the point. The flat side is double-cut and the curved side is single-cut on second-cut and smooth files. This is an extremely useful double-purpose file for flat surfaces and for curved surfaces too large for the round file.

Knife file: The knife file has a wedge-shaped cross-section, the thin edge being straight while the thick edge tapers to the point in approximately the last third of its length. The sides are double-cut. This file is used in filing acute angles.

Dreadnought files: When soft material is being filed, the metal is more readily removed and the teeth of an engineer's file quickly become clogged. When this happens, the file no longer cuts but skids over the surface. This results in constant stoppages to clear the file so that it again cuts properly. To overcome the problem of clogging, files have been developed which have deep curved teeth milled on their faces and these are known as *dreadnought* files, Fig. 2.3.

Figure 2.3 Dreadnought file

These files are designed to remove metal faster and with less effort, since the deep curved teeth produce small spiral filings which clear themselves from the tooth and so prevent clogging. Their principal use is in filing soft materials such as aluminium, lead, white metal, copper, bronze, and brass. They can also be used on large areas of steel, as well as on non-metallic materials such as plastics, wood, fibre, and slate.

This type of file is available as hand, flat, half-round, and square, from 150 mm to 400 mm long. The available cuts are broad, medium, standard, fine, and extra fine.

Needle files: Needle files are used for very fine work in tool making and fitting, where very small amounts of material have to be removed in intricate shapes or in a confined space. This type of file is available from 120 mm to 180 mm long, of which approximately half is file-shaped and cut, the remainder forming a slender circular handle, Fig. 2.4.

Figure 2.4 Needle file

Filing

One of the greatest difficulties facing the beginner is to produce a filed surface which is flat. By carefully observing a few basic principles and carrying out a few exercises, the beginner should be able to produce a flat surface.

Filing is a two-handed operation, and the first stage is to grip the file correctly. The handle is gripped in the palm of the right hand with the thumb on top and the palm of the left hand resting at the point of the file. Having gripped the file correctly, the second stage is to stand correctly at the vice. The left foot is placed well forward to take the weight of the body on the forward stroke. The right foot is placed well back to enable the body to be pushed forward.

Remember that the file cuts on the forward stroke and therefore the pressure is applied by the left hand during the forward movement and is released coming back. Do not lift the file from the work on the back stroke, as the dragging action helps clear the filings from the teeth and also prevents the 'see-saw' action which results in a surface which is curved rather than flat. Above all, take your time – long steady strokes using the length of the file will remove metal faster and produce a flatter surface than short rapid strokes.

Care of files

A file which cuts well saves you extra work. It is important, therefore, that all the teeth are cutting. Never throw files on top of each other in a drawer, as the teeth may be chipped. Never knock the file on its edge to get rid of filings in the teeth – use a file brush. A file brush should be used regularly to remove filings from the teeth, as failure to do so will cause scratching of the work surface and inefficient removal of metal. Always clean the file on completion of the job before putting it away. Do not exert too much pressure when using a new file, or some of the teeth may break off due to their sharpness – work lightly until the fine tooth points are worn slightly. For the same reason, avoid using a new file on rough surfaces of castings, welds, or hard scale.

Always use a properly fitted handle of the correct size – on no account should a file be used without a handle or with a handle which is split; remember, one slip and the tang could pierce your hand.

2.2 The hacksaw

The hacksaw is used to cut metal. Where large amounts of waste metal have to be removed, this is more easily done by hacksawing away the surplus rather than by filing. If the workpiece is left slightly too large, a file can then be used to obtain the final size and surface.

The hacksaw blade fits into a hacksaw frame on two holding pins, one of which is adjustable in order to tension the blade. The hacksaw frame should be rigid, hold the blade in correct alignment, tension the blade easily, and have a comfortable grip.

The blade is fitted to the frame with the teeth pointing away from the handle, Fig. 2.5, and is correctly tensioned by turning the wing nut to take up the slack and then applying a further three turns only. A loose blade will twist or buckle and not cut straight, while an overtightened blade could pull out the ends of the blade.

The standard hacksaw blade is 300 mm long × 13 mm wide × 0.65 mm thick and is available with 14, 18, 24, and 32 teeth per 25 mm; i.e. for every 25 mm length of blade there are 14 teeth, 18 teeth, and so on.

A hacksaw blade should be chosen to suit the type of material being cut, whether hard or soft, and the nature of the cut, whether thick section or thin. Two important factors in the choice of a blade are the pitch, or distance between each tooth, and the material from which the blade is made.

When cutting soft metals, more material will be cut on each stroke and this material must have somewhere to go. The only place the material can go is between the teeth, and therefore if the teeth are further apart there is more space for the metal being cut. The largest space is in the blade having the least number of teeth, i.e. 14 teeth per 25 mm. The opposite is true when cutting harder metals. Less material will be removed on each stroke, which will require less space between each tooth. If less space is required, more teeth can be put in the blade, more teeth are cutting, and the time and effort in cutting will be less.

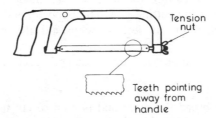

Tension nut

Teeth pointing away from handle

Figure 2.5 Hacksaw

When cutting thin sections such as plate, at least three consecutive teeth must always be in contact with the metal or the teeth will straddle the thin section. The teeth will therefore have to be closer together, which means more teeth in the blade, i.e. 32 teeth per 25 mm.

Like a file, the hacksaw cuts on the forward stroke, which is when pressure should be applied. Pressure should be released on the return stroke. Do not rush but use long steady strokes (around 70 strokes per minute when using high-speed-steel blades). The same balanced stance should be used as for filing.

Table 2.1 gives recommendations for the number of teeth per 25 mm on blades used for hard and soft materials of varying thickness.

Table 2.1 Selection of hacksaw blades

Material thickness (mm)	No. of teeth per 25 mm	
	Hard materials	Soft materials
Up to 3	32	32
3 to 6	24	24
6 to 13	24	18
13 to 25	18	14

Three types of hacksaw blade are available: all-hard, flexible, and bimetal.

- *All hard* – this type is made from hardened high-speed steel. Due to their all-through hardness, these blades have a long blade life but are also very brittle and are easily broken if twisted during sawing. For this reason they are best suited to the skilled user.
- *Flexible* – this type of blade is also made from high-speed steel, but with only the teeth hardened. This results in a flexible blade with hard teeth which is virtually unbreakable and can therefore be used by the less experienced user or when sawing in an awkward position. The blade life is reduced due to the problem of fully hardening the teeth only.
- *Bimetallic* – this type of blade consists of a narrow cutting-edge strip of hardened high-speed steel joined to a tough alloy-steel back by electron beam welding. This blade combines the qualities of hardness of the all-hard blade and the unbreakable qualities of the flexible blade, resulting in a shatterproof blade with long life and fast-cutting properties.

2.3 Cold chisels

Cold chisels are used for cutting metal. They are made from high-carbon steel, hardened and tempered at the cutting end. The opposite end, which is struck by the hammer, is not hardened but is left to withstand the hammer blows without chipping.

Cold chisels are classified as 'flat' or 'cross-cut', according to the shape of the point.

Flat: This chisel has a broad flat point and is used to cut thin sheet metal, remove rivet heads, or split corroded nuts. The cutting edge is ground to an angle of approximately 60°, Fig. 2.6.

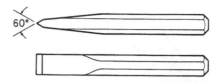

Figure 2.6 'Flat' cold chisel

Cross-cut: This chisel has a narrower point than the flat chisel and is used to cut keyways, narrow grooves, square corners, and holes in sheet metal too small for the flat chisel, Fig. 2.7.

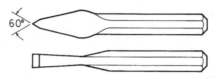

Figure 2.7 'Cross-cut' cold chisel

Using the chisel

When using a cold chisel on sheet-material, great care must be taken not to distort the metal. To prevent distortion, the sheet must be properly supported. A small sheet is best held in a vice, Fig. 2.8. A large sheet can be supported by using two metal bars securely clamped, Fig. 2.9.

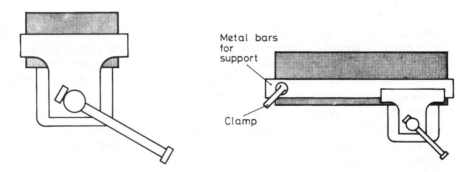

Figure 2.8 Sheet metal in vice   **Figure 2.9** Sheet metal in support bars

To remove a section from the centre of a plate, the plate can be supported on soft metal. It is best to mark out the shape required, drill a series of holes in the waste material, and use the chisel to break through between the holes, Fig. 2.10.

The chisel should be held firmly but not too tight, and the head should be struck with sharp blows from the hammer, keeping your eye on the cutting edge, not the chisel head. Hold the chisel at approximately 40°, Fig. 2.11. Do not hold the chisel at too steep an angle, otherwise it will tend to dig into the metal. Too shallow an angle will cause the chisel to skid and prevent it cutting. Use a hammer large enough to do the job, grasping it well back at the end of the handle, not at the end nearest the head. Never allow a large 'mushroom' head to form on the head of a chisel, as a glancing blow from the hammer can dislodge a chip which could fly off and

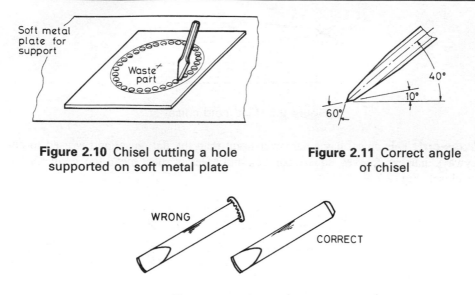

Figure 2.10 Chisel cutting a hole supported on soft metal plate

Figure 2.11 Correct angle of chisel

Figure 2.12 Correct chisel

damage your eye or hand. Always grind off any sign of a mushroom head as it develops, Fig. 2.12.

Cold chisels can be sharpened by regrinding the edge on an off-hand grinder. When resharpening, do not allow the chisel edge to become too hot, otherwise it will be tempered, lose its hardness, and be unable to cut metal.

2.4 Scrapers

Scraping, unlike filing or chiselling, is not done to remove a great deal of material. The material is removed selectively in small amounts, usually to give a flat or a good bearing surface. A surface produced by machining or filing may not be good enough as a bearing where two surfaces are sliding or rotating. The purpose of scraping is therefore to remove high spots to make the surface flat or circular, and at the same time to create small pockets in which lubricant can be held between the two surfaces. Surface plates and surface tables are examples of scraping being used when flatness is of prime importance. Examples where both flatness and lubricating properties are required can be seen on the sliding surfaces of centre lathes and milling, shaping, and grinding machines.

The flat scraper, for use on flat surfaces, resembles a hand file thinned down at the point, but it does not have any teeth cut on it, Fig. 2.13. The point is slightly curved, and the cutting edges are kept sharp by means of an oilstone. The scraper cuts on the forward stroke, the high spots being removed one at a time by short forward strokes. The flatness is checked with reference to a surface plate. A light film of engineer's blue is smeared evenly on the surface plate, and the surface being scraped is placed on top and moved slightly from side to side. Any high spots show up as blue spots, and these are reduced by scraping. The surface is again checked, rescraped, and the process is repeated until the desired flatness is obtained. Flatness of the surface is indicated when the whole area being scraped is evenly covered by blue from the surface plate.

The same procedure is used on internal curved surfaces, using a half-round scraper slightly hollow on the underside, to prevent digging in, and with a cutting edge on

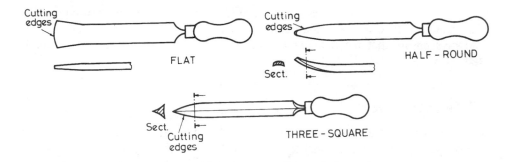

Figure 2.13 Scrapers

each side, Fig. 2.13. The reference surface in this case is the shaft which is to run in the curved surface and which is smeared with engineer's blue. Entry of the shaft in the bearing indicates the high spots, which are removed by scraping, and this process is repeated until the desired surface is produced.

The three-square or triangular scraper, Fig. 2.13 is commonly used to remove the sharp edges from curved surfaces and holes. It is not suited to scraping internal curved surfaces, due to the steeper angle of the cutting edges tending to dig into the surface. However, the sharp point is useful where a curved surface is required up to a sharp corner.

2.5 Engineer's hammers

The engineer's hammer consists of a hardened and tempered steel head, varying in mass from 0.1 kg to about 1 kg, firmly fixed on a tough wooden handle, usually hickory or ash.

The flat striking surface is known as the face, and the opposite end is called the pein. The most commonly used is the ball-pein, Fig. 2.14, which has a hemispherical end and is used for riveting over the ends of pins and rivets.

For use with soft metal such as aluminium or with finished components where the workpiece could be damaged if struck by an engineer's hammer, a range of hammers is available with soft faces, usually hide, copper, or a tough plastic such as nylon. The soft faces are usually in the form of replaceable inserts screwed into the head or into a recess in the face, Fig. 2.15.

Always use a hammer which is heavy enough to deliver the required force but not too heavy to be tiring in use. The small masses, 0.1 kg to 0.2 kg, are used for centre punching, while the 1 kg ones are used with large chisels or when driving large keys or collars on shafts. The length of handle is designed for the appropriate head mass, and the hammer should be gripped near the end of the handle to deliver the required

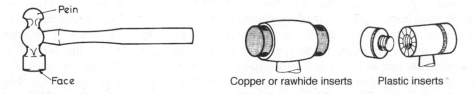

Figure 2.14 Ball-pein hammer   **Figure 2.15** Soft-faced hammers

blow. To be effective, a solid sharp blow should be delivered and this cannot be done if the handle is held too near the hammer head.

Always ensure that the hammer handle is sound and that the head is securely fixed.

2.6 Screwdrivers

The screwdriver is one of the most common tools, and is also the one most misused. Screwdrivers should be used only to tighten or loosen screws. They should never be used to chisel,open tins, scrape off paint, or lever off tight parts such as collars on shafts. Once a screwdriver blade, which is made from toughened alloy steel, has been bent, it is very difficult to keep it in the screw head.

There are two types of screw slot: the straight slot and the cross slot, i.e. 'pozidriv' or 'supadriv'. Always select the screwdriver to suit the size of screw head and the type of slot – use of the incorrect size or type results in damage to both the screwdriver blade and the screw head and in a screw very difficult to loosen or tighten. Cross-slot sizes are numbered, 1, 2, 3, and 4, and screwdrivers are available with corresponding point sizes to suit 'pozidriv' and supadriv' slots.

Straight slots in screws are machined with parallel sides. It is essential that any screwdriver used in such a slot has the sides of the blade parallel to slightly tapered up to about 10°, Fig. 2.16(a). A screwdriver sharpened to a point like a chisel will not locate correctly and will require great force to keep it in the slot, Fig. 2.16(b). Various blade lengths are available with corresponding width and thickness to suit the screw size.

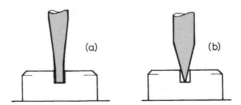

Figure 2.16 Screwdriver point: (a) right, (b) wrong

In the interests of personal safety, never hold the work in your hand while tightening or loosening a screw – the blade may slip and cause a nasty injury. Always hold the work securely in a vice or clamped to a solid surface.

2.7 Taps

Tapping is the operation of cutting an internal thread by means of a cutting tool known as a tap. When tapping by hand, straight-flute hand taps are used. These are made from hardened high-speed steel and are supplied in sets of three. The three taps differ in the length of chamfer at the point, known as the lead. The one with the longest lead is referred to as the taper or first tap, the next as the second or intermediate tap, and the third, which has a very short lead, as the bottoming or plug tap, Fig. 2.17 A square is provided at one end so that the tap can be easily rotated by holding it in a tap wrench, Fig. 2.18. The chuck type of wrench is used for the smaller tap sizes.

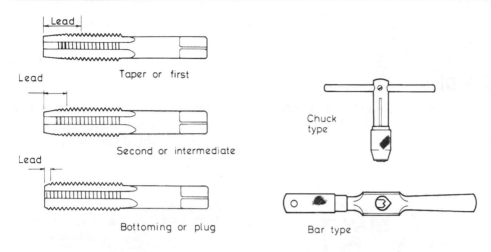

Figure 2.17 Set of taps   **Figure 2.18** Tap wrenches

The first stage in tapping is to drill a hole of the correct size. This is known as the tapping size and is normally slightly larger than the root diameter of the thread. Table 2.2 shows the tapping sizes for ISO metric threads which have replaced most threads previously used in Great Britain.

Tapping is then started using the taper or first tap securely held in a tap wrench. The long lead enables it to follow the drilled hole and keep square. The tap is rotated, applying downward pressure until cutting starts. No further pressure is required, since the tap will then screw itself into the hole. The tap should be turned back quite often, to help clear chips from the flutes.

If the hole being tapped passes through the component, it is only necessary to repeat the operation using the second or intermediate tap. Where the hole does not pass through – known as a blind hole – it is necessary to use the plug or bottoming tap. This tap has a short lead and therefore forms threads very close to the bottom of the hole. When tapping a blind hole, great care should be taken not to break the tap. The tap should be occasionally withdrawn completely and any chips be removed before proceeding to the final depth.

Table 2.2 Tapping sizes for ISO metric threads

Thread diameter and pitch (mm)	Drill diameter for tapping (mm)
1.6×0.35	1.25
2×0.4	1.6
2.5×0.45	2.05
3×0.5	2.5
4×0.7	3.3
5×0.8	4.2
6×1.0	5.0
8×1.25	6.8
10×1.5	8.5
12×1.75	10.2

For easier cutting and the production of good-quality threads, a proprietary tapping compound should be used.

2.8 Dies

Dies are used to cut external threads and are available in sizes up to approximately 36 mm thread diameter. The common type, for use by hand, is the circular split die, made from high-speed steel hardened and tempered and split at one side to enable small adjustments of size to be made, Fig. 2.19.

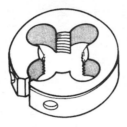

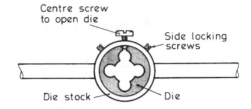

Figure 2.19 Circular split die   **Figure 2.20** Die holder

The die is held in a holder known as a die stock, which has a central screw for adjusting the size and two side locking screws which lock in dimples in the outside diameter of the die, Fig. 2.20. The die is inserted in the holder with the split lined up with the central screw. The central screw is then tightened so that the die is expanded, and the two side locking screws are tightened to hold the die in position.

Dies have a lead on the first two or three threads, to help start cutting, but it is usual also to have a chamfer on the end of the component. The die is placed squarely on the end of the bar and is rotated, applying downward pressure until cutting starts, ensuring that the stock is horizontal. No further pressure is required, since the die then screws itself forward as cutting proceeds. The die should be rotated backwards every two or three turns, to break up and clear the chips. The thread can now be checked with a nut. If it is found to be tight, the central screw is slackened, the side locking screws are tightened, and the die is run down the thread again. This can be repeated until the final size is reached.

As with tapping, easier cutting and better threads are produced when a proprietary cutting compound is used.

2.9 Powered hand tools

The main advantages of powered hand tools are the reduction of manual effort and the speeding up of the operation. The operator, being less fatigued, is able to carry out the task more efficiently, and the speeding up of the operation results in lower production costs. Being portable, a powered hand tool can be taken to the work, which can also lead to a reduction in production costs. Accuracy of metal-removal operations is not as good with powered hand tools, since it is difficult to remove metal from small areas selectively. A comparison of hand and powered hand tools is shown in Table 2.3.

Table 2.3 Comparison of hand and powered hand tools

	Speed of production	Cost of tool	Accuracy	Fatigue
Hand tools	Low	Low	High	High
Powered hand tools	High	High	Low	Low

Powered hand tools can be electric or air-operated. In general, electric tools are heavier than the equivalent air tool, due to their built-in motor, e.g. electric screwdrivers weigh 2 kg while an equivalent air-operated screwdriver weighs 0.9 kg. The cost of powered tools is much greater than the equivalent hand tools and must be taken into account when a choice has to be made.

Air-operated tools can be safely used in most work conditions, while electrical tools should not be used in conditions which are wet or damp or where there is a risk of fire or explosion, such as in flammable or dusty atmospheres. A selection of air-operated tools is shown in Fig. 2.21.

Hand drills
Electric and air-operated drills are available with a maximum drilling capacity in steel of about 30 mm diameter for electric and about 10 mm diameter for air models. Air-operated tools are more ideally suited to the rapid drilling of the smaller diameter holes, Fig. 2.21(a).

Screwdriver
Used for inserting screws of all types, including machine, self-tapping, self-drilling and tapping, and wood screws. Some models are reversible and can be used with equal ease to remove screws. The tool bits are interchangeable to suit the different screw-head types, such as slotted, 'supadriv', 'pozidriv', hexagon-socket, or hexagon-headed. Electric and air-operated screwdrivers are available with a maximum capacity of about 8 mm diameter thread with a variety of torque settings to prevent the screw being overtightened or sheared off, Fig. 2.21(b).

Impact wrench
Used for tightening and also, with the reversal mechanism, for loosening hexagon-headed nuts and screws. Air-powered models are available with a maximum capacity of 32 mm diameter threads and with torque settings to suit a range of thread sizes. They have the advantage of being able to tighten all nuts or screws to the same predetermined load, Fig. 2.21(c).

Grinder
Used to remove metal from the rough surfaces of forgings, castings, and welds, usually when the metal is too hard or the amount to be removed is too great for a file or a chisel. Electric and air-operated grinders are available with straight grinding wheels up to 230 mm diameter or with small mounted points of various shapes and sizes, Fig. 2.21(d).

Metal shears
Used to cut metal, particularly where the sheet cannot be taken to a guillotine or where profiles have to be cut. Electric or air-operated shears are available capable of

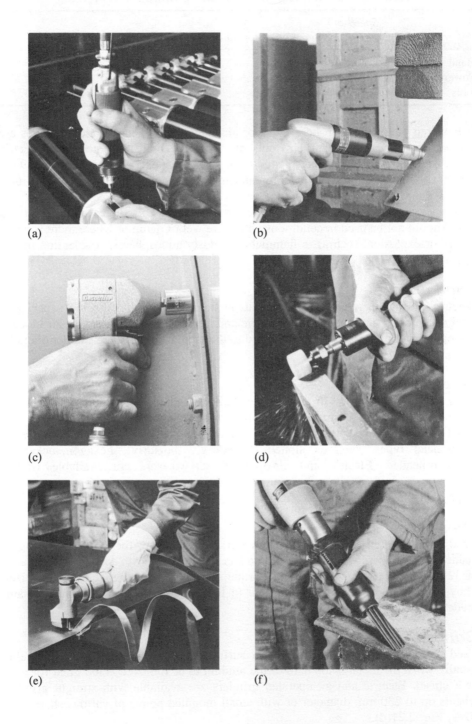

Figure 2.21 Air-operated tools: (a) hand drill, (b) screwdriver, (c) impact wrench, (d) grinder, (e) metal shears, (f) hammer

cutting steel sheet up to 2 mm thick by means of the scissor-like action of a reciprocating blade, Fig. 2.21(e).

Hammer

Can be fitted with a wide range of attachments for riveting, shearing off rivet heads, removing scale, or panel cutting. Air-operated models are available which deliver between 3000 and 4000 blows per minute, Fig. 2.21(f).

3 Marking out

Marking out is the scratching of lines on the surface of a workpiece, known as scribing, and is usually carried out only on a single workpiece or a small number of workpieces. The two main purposes of marking out are:

- to indicate the workpiece outline or the position of holes, slots, etc. If the excess material will have to be removed, a guide is given for the extent to which hacksawing or filing can be carried out;
- to provide a guide to setting up the workpiece on a machine. The workpiece is set up relative to the marking out and is then machined. This is especially important when a datum has to be established when castings and forgings are to be machined.

It is important to note that the scribed lines are only a guide, and any accurate dimension must be finally checked by measuring.

3.1 Datum

The function of a datum is to establish a reference position from which all dimensions are taken and hence all measurements are made. The datum may be a point, an edge or a centre line, depending on the shape of the workpiece. For any plane surface, two datums are required to position a point and these are usually at right angles to each other.

Figure 3.1 shows a workpiece where the datum is a point; Fig. 3.2 shows a workpiece where both datums are edges; Fig. 3.3 shows a workpiece where both datums are centre lines; and Fig. 3.4 shows a workpiece where one datum is an edge and the other is a centre line.

The datums are established by the draughtsman when the drawing is being dimensioned and, since marking out is merely transferring drawing dimensions to the workpiece, the same datums are used.

3.2 Co-ordinates

The draughtsman can dimension drawings in one of two ways.

- *Rectangular co-ordinates* – where the dimensions are taken relative to the datums at right angles to each other, i.e. the general pattern is rectangular. This is the method shown in Figs 3.2 and 3.4.
- *Polar co-ordinates* – where the dimension is measured along a radial line from the datum. This is shown in Fig. 3.1. Marking out polar co-ordinates requires not only accuracy of the dimension along the radial line but accuracy of the angle itself. As the polar distance increases, any slight angular error will effectively increase the inaccuracy of the final position.

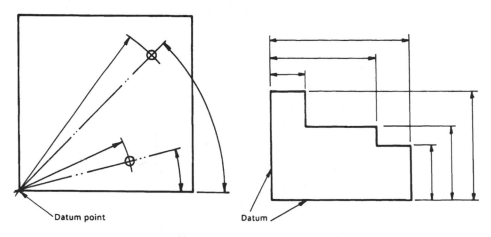

Figure 3.1 Datum point

Figure 3.2 Datum edges

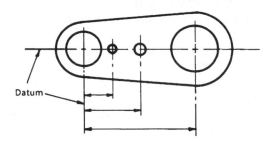

Figure 3.3 Datum centre lines

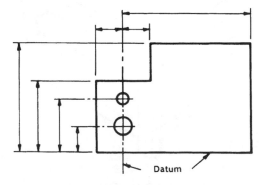

Figure 3.4 Datum edge and centre line

The possibility of error is less with rectangular co-ordinates, and the polar co-ordinate dimensions shown in Fig. 3.1 could be redrawn as rectangular co-ordinates as shown in Fig. 3.5.

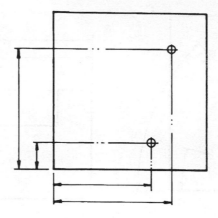

Figure 3.5 Rectangular co-ordinates

3.3 Marking out equipment

Surface table and surface plate

In order to establish a datum from which all measurements are made a reference surface is required. This reference surface takes the form of a large flat surface called a surface table (Fig. 3.6) upon which the measuring equipment is used.

Surface plates (Fig. 3.7) are smaller reference surfaces and are placed on a bench for use with smaller workpieces. For general use, both surface tables and surface plates are made from cast iron machined to various grades of accuracy. For high-accuracy inspection work and for use in standards rooms, surface tables and plates made from granite are available.

Figure 3.6 Surface table

Parallels (Fig. 3.7)

The workpiece can be set on parallels to raise it off the reference surface and still maintain parallelism. Parallels are made in pairs to precisely the same dimensions,

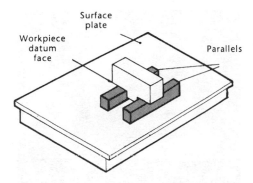

Figure 3.7 Surface plate and parallels

from hardened steel, finish ground, with their opposite faces parallel and adjacent faces square. A variety of sizes should be available for use when marking out.

Jacks and wedges

When a forging or casting, has to be marked out, which has an uneven surface or is awkward in shape, it is still essential to maintain the datum relative to the reference surface. Uneven surfaces can be prevented from rocking and kept on a parallel plane by slipping in thin steel or wooden wedges (Fig. 3.8) at appropriate positions. Awkward shapes can be kept in the correct position by support from adjustable jacks (see Fig. 3.9).

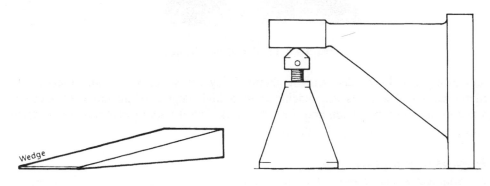

Figure 3.8 Wedge   **Figure 3.9** Jack used to support

Angle plate

When the workpiece has to be positioned at 90° to the reference surface, it can be clamped to an angle plate (Fig. 3.10). Angle plates are usually made from cast iron and the edges and faces are accurately machined flat, square and parallel. Slots are provided in the faces for easy clamping of the workpiece. Angle plates may be plain or adjustable.

Vee blocks (Fig. 3.11)

Holding circular work for marking out or machining can be simplified by using a vee block. The larger sizes are made from cast iron, the smaller sizes from steel hardened

Figure 3.10 Angle plate and surface gauge

Figure 3.11 Vee block in use

and ground, and provided with a clamp. They are supplied in pairs marked for identification. The faces are machined to a high degree of accuracy of flatness, squareness, and parallelism, and the 90° vee is central with respect to the side faces and parallel to the base and side faces.

Engineer's square (Fig. 3.12)
An engineer's square is used when setting the workpiece square to the reference surface (see Fig. 3.13) or when scribing lines square to the datum edge (Fig. 3.14). The square consists of a stock and blade made from hardened steel and ground on all faces and edges to give a high degree of accuracy in straightness, parallelism, and squareness. It is available in a variety of blade lengths.

Combination set (Fig. 3.15)
The combination set consists of a graduated hardened steel rule on which any of three separate heads – protractor, square, or centre head – can be mounted. The rule has a slot in which each head slides and can be locked at any position along its length.

Protractor head (Fig. 3.16): This head is graduated from 0 to 180°, is adjustable through this range, and is used when scribing lines at an angle to a workpiece datum.

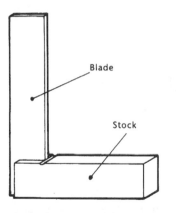

Blade

Stock

Figure 3.12 Engineer's square

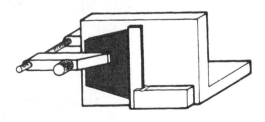

Figure 3.13 Setting workpiece square to reference surface

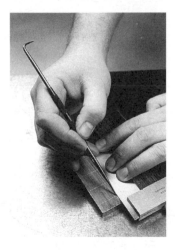

Figure 3.14 Scribing line square to datum

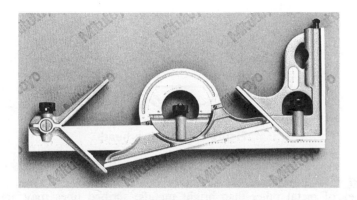

Figure 3.15 Combination set

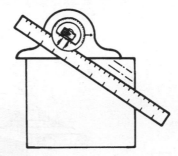

Figure 3.16 Protractor head

Square head (Fig. 3.17(a)): This head is used in the same way as an engineer's square, but, because the rule is adjustable, it is not as accurate. A second face is provided at 45° (Fig. 3.17(b)). A spirit level is incorporated which is useful when setting workpieces such as castings level with the reference surface. Turned on end, this head can also be used as a depth gauge (see Fig. 3.17(c)).

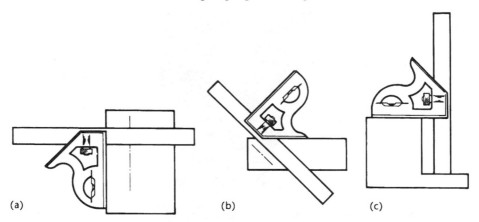

(a)   (b)   (c)

Figure 3.17 Square head

Centre head (Fig. 3.18): With this head the blade passes through the centre of the vee and is used to mark out the centre of a circular workpiece or round bar.

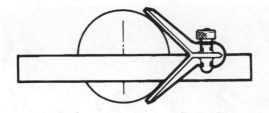

Figure 3.18 Centre head

Marking dye

On surfaces of metal other than bright metals, scribed lines may not be clearly visible. In such cases the surface can be brushed or sprayed with a quick drying coloured dye before marking out. This provides a good contrast, making the scribed lines easy to see.

Scriber (Fig. 3.19)

The scriber is used to scribe all lines on a metal surface and is made from hardened and tempered steel, ground to a fine point which should always be kept sharp to give well-defined lines. Another type is shown in Fig. 3.14.

Figure 3.19 Scriber

Surface gauge (Fig. 3.10)

The surface gauge, also known as a scribing block, is used in conjunction with a scriber to mark out lines on the workpiece parallel with the reference surface. The height of the scriber is adjustable and is set in conjunction with a steel rule. The expected accuracy from this set up will be around 0.3 mm but with care this can be improved.

Vernier height gauge (Fig. 3.20)

Where greater accuracy is required than can be achieved using a surface gauge, marking out can be done using a vernier height gauge. The vernier scale carries a jaw upon which various attachments can be clamped. When marking out, a chisel pointed scribing blade is fitted. Care should be taken to allow for the thickness of the jaw, depending on whether the scribing blade is clamped on top or under the jaw. The precise thickness of the jaw is marked on each instrument. These instruments can be read to an accuracy of 0.02 mm and are available in a range of capacities reading from 0 to 1000 mm.

Dividers and trammels

Dividers are used to scribe circles or arcs and to mark off a series of lengths such as hole centres. They are of spring bow construction, each of the two pointed steel legs being hardened and ground to a fine point and capable of scribing a maximum circle of around 150 mm diameter (Fig. 3.21). Larger circles can be scribed using trammels, where the scribing points are adjustable along the length of a beam (Fig. 3.22).

Dividers and trammels are both set in conjunction with a steel rule by placing one point in a convenient graduation line and adjusting the other to coincide with the graduation line the correct distance away.

Hermaphrodite calipers (Fig. 3.23)

These combine a straight pointed divider leg with a caliper or stepped leg and are used to scribe a line parallel to the edge of a workpiece. They are more commonly known as 'odd-legs' or 'jennies'.

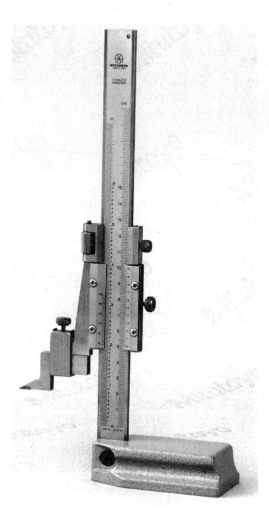

Figure 3.20 Vernier height gauge

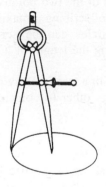

Figure 3.21 Dividers

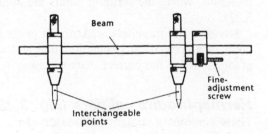

Beam

Fine-adjustment screw

Interchangeable points

Figure 3.22 Trammels

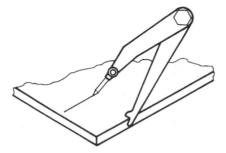

Figure 3.23 Hermaphrodite calipers

Precision steel rule

These are made from hardened and tempered stainless steel, photoetched for extreme accuracy and have a non-glare satin chrome finish. Rules are available in lengths of 150 mm and 300 mm and graduations may be along each edge of both faces usually in millimetres and half millimetres.

Accuracy of measurement depends on the quality of the rule and the skill of the operator. The width of the lines on a high-quality rule are quite fine and accuracies of around 0.15 mm can be achieved but an accuracy of double this can more realistically be achieved.

Centre punch (Fig. 3.24)

The centre punch is used to provide a centre location for dividers and trammels when scribing circles or arcs, or to show permanently the position of a scribed line by a row of centre dots. The centre dot is also used as a start for small diameter drills.

Centre punches are made from high carbon steel, hardened and tempered with the point ground at 30° when used to provide a centre location for dividers and at 90° for other purposes.

Care should be taken in the use of centre dots on surfaces which are to remain after machining, since, depending upon the depth, they may prove difficult to remove.

Figure 3.24 Centre punch

Clamps

Clamps are used when the workpiece has to be securely fixed to another piece of equipment, e.g. to the face of an angle plate (Fig. 3.10).

The type most used are toolmaker's clamps (Fig. 3.25), which are adjustable within a range of about 100 mm but will only clamp parallel surfaces. Greater thicknesses can be clamped using 'G' clamps, so named because of their shape (Fig. 3.26). Due to the swivel pad on the end of the clamping screw, the'G' clamp is also capable of clamping surfaces which are not parallel.

Care should be taken to avoid damage to the surfaces by the clamp.

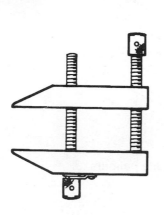

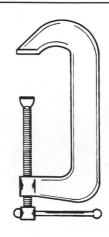

Figure 3.25 Toolmaker's clamp   **Figure 3.26** 'G' clamp

3.4 Examples of marking out

We will now see how to mark out a number of components.

] *Example 3.1: component shown in Fig. 3.27*

The plate shown at step 1 has been filed to length and width with the edges square and requires the position of the steps to be marked out.

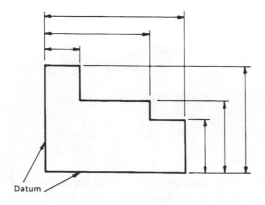

Datum

Figure 3.27 Component used in Example 3.1

Step 1: Use a square on one datum edge and measure the distance from the other datum edge using a precision steel rule. Scribe lines.

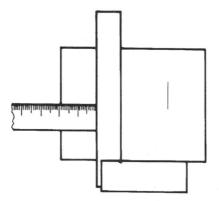

Step 2: Repeat with the square on the second datum edge and scribe lines to intersect.

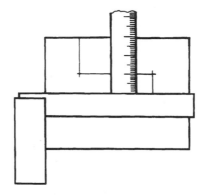

Example 3.2: component shown in Fig. 3.28

The plate shown at step 1 has been cut out 2 mm oversize on length and width and has not been filed. All four sides have sawn edges.

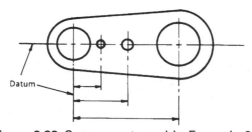

Figure 3.28 Component used in Example 3.2

Step 1: Measure from each long edge and find the centre using a precision steel rule. Scribe the centre line using the edge of the rule as a guide. Find the centre of the small radius by measuring from one end the size of the radius plus 1 mm (this allows for the extra left on the end). Centre dot where the lines intersect.

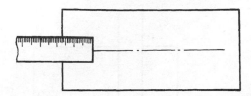

Step 2: Using dividers, set the distance from the centre of the small radius to the centre of the first small hole. Scribe an arc. Repeat for the second small hole and the large radius. Centre dot at the intersection of the centre lines. The dividers are set using the graduations of a precision steel rule.

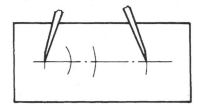

Step 3: Set dividers to the small radius. Locate on the centre dot and scribe the radius. Repeat for the large radius and if necessary, the two holes.

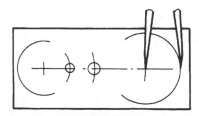

Step 4: Complete the profile by scribing a line tangential to the two radii using the edge of a precision steel rule as a guide.

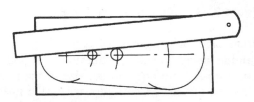

Example 3.3: component shown in Fig. 3.29

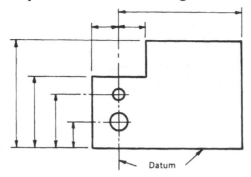

Figure 3.29 Component used in Example 3.3

The plate shown at step 1 has been roughly cut to size and requires complete marking out of the profile and holes.

Step 1: Clamp the plate to the face of an angle plate, ensuring that the clamps will not interfere with marking out. Use a scriber in a surface gauge and set the heights in conjunction with a precision steel rule. Scribe the datum line. Scribe each horizontal line the correct distance from the datum.

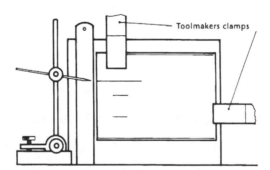

Step 2: Without unclamping the plate, swing the angle plate on to its side (note the importance of clamp positions at step 1). This ensures that the lines about to be scribed are at right angles to those scribed in step 1, owing to the accuracy of the angle plate. Scribe the datum centre line. Scribe each horizontal line the correct distance from the datum to intersect the vertical lines.

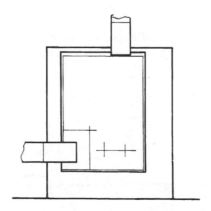

☐ *Example 3.4: component shown in Fig. 3.30*

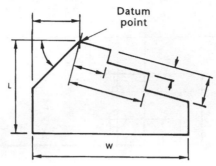

Figure 3.30 Component used in Example 3.4

The plate shown at step 1 is to be produced from the correct width (*W*) bright rolled strip and has been sawn 2 mm oversize on length (*L*). The base edge has been filed square to the ends and it is required to mark out the angled faces.

Step 1: Using a precision steel rule measure from two adjacent edges to determine the datum point. Centre dot the datum point.

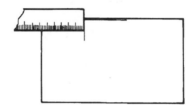

Step 2: Set protractor at required angle and scribe line through datum point.

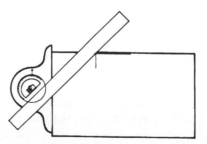

Step 3: Reset protractor at second angle and scribe one line through datum point. Scribe the remaining two lines parallel to and the correct distance from the first line using the protractor at the same setting.

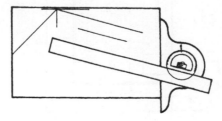

Step 4: Set dividers at correct distances, locate in datum centre dot and mark positions along scribed line.

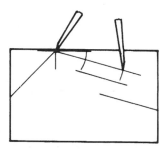

Step 5: Reset protractor and scribe lines through marked positions.

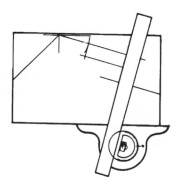

◻ *Example 3.5: shaft shown in Fig. 3.31*

The shaft shown is to have a keyway cut along its centre line for a required length. Accurate machining is made possible by setting up the shaft relative to the marked out position.

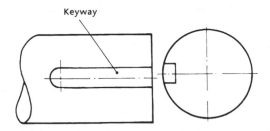

Keyway

Figure 3.31 Component used in Example 3.5

Step 1: Scribe line on the end face through the centre of the shaft using the centre head of a combination set (see Fig. 3.18).

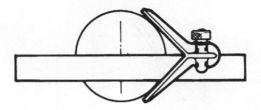

Step 2: Clamp shaft in a vee block ensuring that the line marked at step 1 is lying horizontal. This can be checked using a scriber in a surface gauge. Transfer the centre line along the required length of shaft. Scribe two further lines to indicate the width of slot.

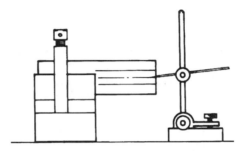

Step 3: The required length of slot can be marked without removing it simply by turning the vee block on its end and scribing a horizontal line at the correct distance from the end of the shaft.

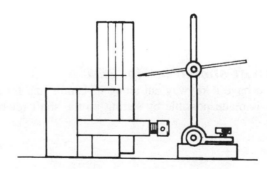

4 Sheet-metal operations

Many engineering components are produced from a flat sheet of metal which is cut to shape and then folded to form the finished article. The edges are then secured by a variety of methods such as welding, brazing, soldering, and riveting. The accuracy of the size and shape of the finished article depends upon the accuracy of drawing the shape on the flat sheet, known as the development. Allowance is made at this stage for folding or bending, the amount varying with the radius of the bend and the metal thickness.

The thickness of metal sheet is identified by a series of numbers known as standard wire gauge, or SWG. Table 4.1 lists the most frequently used gauges and gives their thickness in millimetres.

Table 4.1 Most frequently used standard wire gauges

SWG	Thickness (mm)
10	3.2
12	2.6
14	2.0
16	1.6
18	1.2
20	1.0
22	0.9
24	0.7
	0.6

Cutting is carried out using simple snips for thin-gauge steel up to around 20 SWG, treadle-operated guillotines capable of cutting 14 SWG steel, and hand-lever shears for sheet up to 5 mm thick.

Holes and apertures can be cut using a simple hand-operated punch or a punch and die fitted in a fly press.

For simple bends in thin material, bending can be carried out in a vice. For bends in thicker material with a specific bend radius, folding machines give greater accuracy with less effort.

4.1 Cutting and bending sheet metal

Light-gauge metal can be easily cut using snips. These may have straight or curved blades, Fig. 4.1, the latter being used to cut around a curved profile. Lengths of handle vary from 200 mm to 300 mm, the longer handle giving greater leverage for cutting heavier gauge material. For cutting thicker metals, up to 1.5 mm, hand-lever shears are available, usually bench-mounted, Fig. 4.2. The length of the lever and the linkage to the moving shear blade ensure adequate leverage to cut the thicker metals.

Figure 4.1 Straight- and curved-blade snips

Figure 4.2 Hand-lever shears

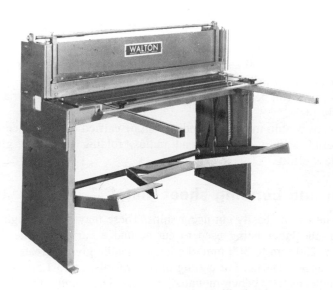

Figure 4.3 Treadle guillotine

Where larger sheets are required to be cut with straight edges, the guillotine is used. Sheet widths of 600 mm × 2 mm thick and up to 1200 mm × 1.6 mm thick can be accommodated in treadle-operated guillotines, Fig. 4.3. These have a moving top blade, which is operated by a foot treadle, and a spring which returns the blade to the top of its stroke. The table is provided with guides, to maintain the cut edges square, and adjustable stops to provide a constant size over a number of components. When the treadle is operated, a clamp descends to hold the work in position while cutting takes place, and this also acts as a guard to prevent injury. *These machines can be extremely dangerous if not used correctly, so take great care.*

When holes are to be cut in sheet metal, this can be done simply and effectively using 'Q-Max' cutters as shown in Fig. 4.4. A pilot hole is drilled in the correct position, the screw is inserted with the punch and die on either side of the sheet, and the screw is tightened. The metal is sheared giving a correct size and shape of hole in the required position.

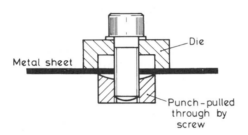

Figure 4.4 'Q-Max' cutter

Where a number of components require the same size hole in the same position, it may be economical to manufacture a punch and die for the operation. The operation is carried out on a fly press, Fig. 4.5, with the punch, which is the size and

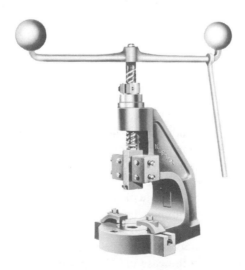

Figure 4.5 Fly press

shape of the hole required, fitted in the moving part of the press. The die, which contains a hole the same shape as the punch, but slightly larger to give clearance, is clamped to the table directly in line with the punch. When the handle of the fly press is rotated, the punch descends and a sheet of metal inserted between the punch and die will have a piece removed the same shape as the punch, Fig. 4.6.

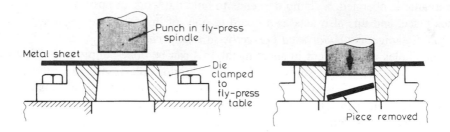

Figure 4.6 Punch and die in fly press

With the use of simple tools, the fly press can also be used for bending small components, Fig. 4.7. The top tool is fixed to the moving part and the bottom tool, correctly positioned under the top tool, is fixed to the table of the press. Metal bent in this way will spring back slightly, and to allow for this the angle of the tool is made less than 90°. In the case of mild steel, an angle of 88° is sufficient for the component to spring back to 90°.

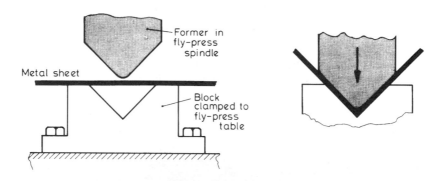

Figure 4.7 Bending tool in fly press

The simplest bends can be produced by holding the component in a vice and bending it over using a soft hammer. If the component is wider than the vice jaws, it can be clamped between metal bars. Unless a radius is put on one of the bars, this method produces a sharp inside corner, which may not always be desirable.

Folding machines, Fig. 4.8, are used with larger work of thicker gauges and for folding box sections. The top clamping beam, is adjustable to allow for various thicknesses of material and can be made up in sections known as fingers to accommodate a previous fold. Slots between the fingers allow a previous fold not to interfere with further folds, as in the case of a box section where four sides have to be folded. The front folding beam, pivoted at each end, is operated by a handle which folds the metal past the clamping blade, Fig. 4.9.

Figure 4.8 Folding machine

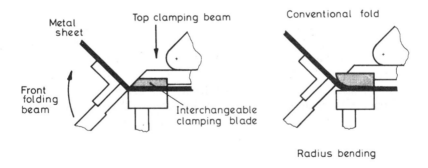

Figure 4.9 Folding operations

4.2 Development

The development of sheet-metal components ranges from the simple to the extremely complex. Let us consider three simple shapes: a cylinder, a cone, and a rectangular tray.

If a cylinder is unfolded, like unrolling a carpet, the length of the development is equal to the circumference, Fig. 4.10.

If a cone is unfolded while pivoting about the apex O, the development is a segment of a circle of radius Oa whose arc ab is equal in length to the circumference of the base, Fig. 4.11. To find the length of arc ab, the base diameter is equally divided into twelve parts. The twelve small arcs 1–2, 2–3, etc. are transferred to the arc with point 12 giving the position of point b. A part cone (frustum) is developed in exactly the same way, with the arc representing the small diameter having a radius Oc, Fig. 4.12.

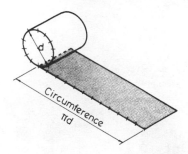

Figure 4.10 Development of cylinder

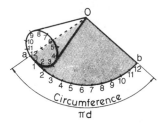

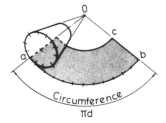

Figure 4.11 Development of cone   **Figure 4.12** Development of part cone

In practice, the circumference must take account of the material thickness. Any metal which is bent will stretch on the outside of the bend and be compressed on the inside. Unless the metal is of very light gauge, an allowance must be made for this. The allowance is calculated on the assumption that, since the outside of the bend stretches and the inside is compressed, the length at a distance half way between the inside and outside diameters, i.e. the mean diameter, will remain unchanged.

Example 4.1

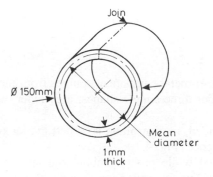

Figure 4.13 Cylinder

The cylinder shown in Fig. 4.13 has an outside diameter of 150 mm and is made from 19 SWG (1 mm thick) sheet. Since the outside diameter is 150 mm and the thickness 1 mm

mean diameter = 150 − 1 = 149 mm
giving
mean circumference = $\pi \times 149 = 468$ mm

The circumference at the outside of the cylinder is

$\pi = 150 = 471$ mm

Thus a blank cut to a length of 468 mm will stretch to a length of 471 mm at the outside and give a component of true 150 mm diameter.

Example 4.2

The development of a rectangular tray is simply the article with the sides and ends folded down, Fig. 4.14(a). The development would be as shown in Fig. 4.14(b), the dotted lines indicating the position of the bend. If sharp inside corners are permissible, the bend lines are the inside dimensions of the tray. If the tray is dimensioned to the outside, you must remember to deduct twice the metal thickness for length and for width.

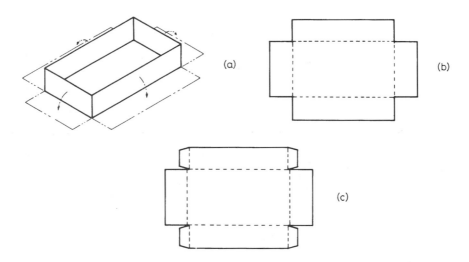

Figure 4.14 Development of rectangular tray

A tray which is to be joined by spot welding or soldering requires a tab, and in this case the tab bend line must allow for the metal thickness so that the tab fits against the inside face of the tray, Fig. 4.14(c). Sharp inside corners for bends are not always possible or desirable, and as a general rule an inside radius is made, equal to twice the thickness of the metal used.

To find the development length on the flat sheet, it is necessary to find the length of the mean line by calculating the lengths of the flat portions and the bends separately. The stretched-out length of the bend is called the bend allowance and for a 90° bend is found by multiplying the mean radius by 1.57 (i.e. $\pi/2$).

☐ *Example 4.3*

Figure 4.15 shows a right-angled bracket made from 1 mm thick material. To obtain the development, first find

$$\text{length } ab = 60 - \text{inside radius} - \text{metal thickness}$$
$$= 60 - 2 - 1$$
$$= 57 \text{ mm}$$

next

$$\text{length } cd = 80 - 2 - 1$$
$$= 77 \text{ mm}$$

finally

$$\text{length } bc = \text{mean radius} \times 1.57$$

$$= (\text{inside radius} + \tfrac{1}{2} \text{ metal thickness}) \times 1.57$$

$$= (2 + 0.5) \times 1.57$$
$$= 2.5 \times 1.57$$
$$= 3.9, \text{ say 4 mm}$$

∴ total length of development $= 57 + 77 + 4 = 138$ mm

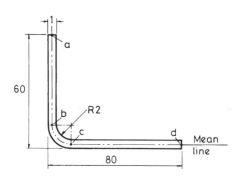

Figure 4.15 Right-angled bracket   **Figure 4.16** Development of right-angled bracket

It can be seen that the development of the bracket shown in Fig. 4.15 is made up of a 57 mm straight length plus a bend allowance of 4 mm plus a further straight length of 77 mm, as shown in Fig. 4.16. The bend is half way across the bend allowance, and therefore the bend line must be $57 + 2 = 59$ mm from one edge.

5 Measuring equipment

Some form of precise measurement is necessary if parts are to fit together as intended no matter whether the parts were made by the same person, in the same factory, or in factories a long way apart. Spare parts can then be obtained with the knowledge that they will fit a part which was perhaps produced years before.

To achieve any degree of precision, the measuring equipment used must be precisely manufactured with reference to the same standard of length. That standard is the metre, which is now defined in terms of the wavelength of a particular light. Having produced the measuring equipment to a high degree of accuracy, it must be used correctly. You must be able to assess the correctness of size of the work by adopting a sensitive touch or 'feel' between the instrument and work. This 'feel' can be developed only from experience of using the instrument, although some instruments do have an aid such as the ratchet stop on some micrometers. Having the correct equipment and having developed a 'feel', you must be capable of reading the instrument to determine the workpiece size. It is here that the two main types of length-measuring instrument differ: the micrometer indicates the linear movement of a rotating precision screw thread, while the vernier instruments compare two scales which have a small difference in length between their respective divisions.

5.1 Vernier instruments

All instruments employing a vernier consist of two scales: one moving and one fixed. The fixed scale is graduated in millimetres, every 10 divisions equalling 10 mm, and is numbered 0, 1, 2, 3, 4 up to the capacity of the instrument. The moving or vernier scale is divided into 50 equal parts which occupy the same length as 49 divisions or 49 mm on the fixed scale (see Fig. 5.1). This means that the distance between each graduation on the vernier scale is $^{49}\!/_{50}$ mm $= 0.98$ mm, or 0.02 mm less than each division on the fixed scale (see Fig. 5.2(a)).

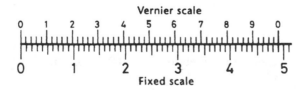

Figure 5.1 Vernier scale

If the two scales initially have their zeros in line and the vernier scale is then moved so that its first graduation is lined up with a graduation on the fixed scale, the zero on the vernier scale will have moved 0.02 mm (Fig. 5.2(b)). If the second graduation is lined up, the zero on the vernier scale will have moved 0.04 mm (Fig. 5.2(c)), and so on. If graduation 50 is lined up, the zero will have moved $50 \times 0.02 = 1$ mm.

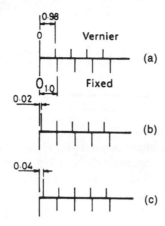

Figure 5.2 Vernier scale readings

Since each division on the vernier scale represents 0.02 mm, five divisions represent $5 \times 0.02 = 0.1$ mm. Every fifth division on this scale is marked 1 representing 0.1 mm, 2 representing 0.2 mm, and so on (Fig. 5.1).

To take a reading, note how many millimetres the zero on the vernier scale is from zero on the fixed scale. Then note the number of divisions on the vernier scale from zero to a line which exactly coincides with a line on the fixed scale.

In the reading shown in Fig. 5.3(a) the vernier scale has moved 40 mm to the right. The eleventh line coincides with a line on the fixed scale, therefore $11 \times 0.02 = 0.22$ mm is added to the reading on the fixed scale, giving a total reading of 40.22 mm.

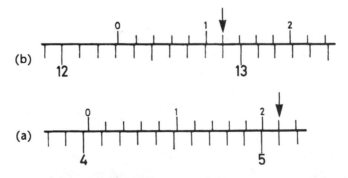

Figure 5.3 Vernier readings

Similarly, in Fig. 5.3(b) the vernier scale has moved 120 mm to the right plus 3 mm and the sixth line coincides, therefore, $6 \times 0.02 = 0.12$ mm is added to 123 mm, giving a total of 123.12 mm.

It follows that if one part of a measuring instrument is attached to the fixed scale and another part to the moving scale, we have an instrument capable of measuring to 0.02 mm.

Vernier caliper

The most common instrument using the above principle is the vernier caliper (see Fig. 5.4). These instruments are capable of external, internal, step and depth measurements (Fig. 5.5) and are available in a range of measuring capacities from 150 mm to 1000 mm.

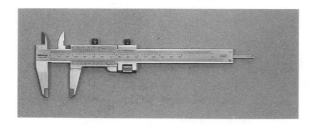

Figure 5.4 Vernier caliper

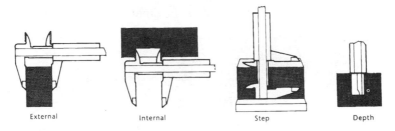

External   Internal   Step   Depth

Figure 5.5 External, internal, step and depth measurement

To take a measurement, slacken both locking screws A and B (Fig. 5.6). Move the sliding jaw along the beam until it contacts the surface of the work being measured. Tighten locking screw B. Adjust the nut C until the correct 'feel' is obtained, then tighten locking screw A. Re-check 'feel' to ensure that nothing has moved. When you are satisfied, take the reading on the instrument.

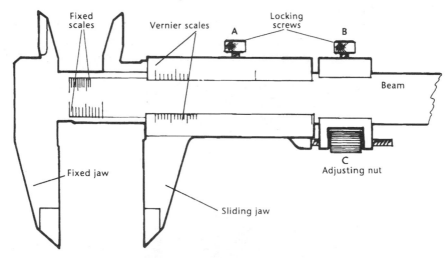

Figure 5.6 Vernier caliper adjustment

Dial calipers (Fig. 5.7) are a form of vernier caliper where readings of 1 mm steps are taken from the vernier beam and sub-divisions of this are read direct on a dial graduated in 0.02 mm divisions.

The electronic caliper shown in Fig. 5.8 is a modern-designed precision measuring instrument which has an electronic measuring unit with an LCD digital readout giving direct readings in imperial or metric units with a resolution of 0.0005" or 0.01 mm.

Figure 5.7 Dial caliper

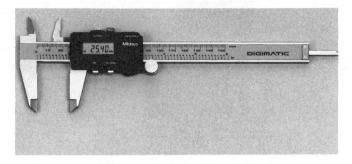

Figure 5.8 Electronic caliper

Vernier height gauge

The above principles apply to the vernier height gauge, Fig. 5.9. In this case the beam, carrying the fixed scale, is attached to a heavy base. The vernier scale carries a jaw upon which various attachments can be clamped. It is most widely used with

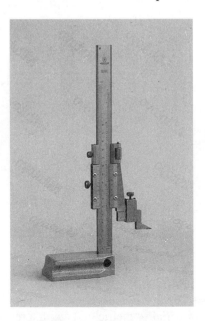

Figure 5.9 Vernier height gauge

a chisel-pointed scribing blade for accurate marking out, as well as for checking the height of steps in components. Care should be taken to allow for the thickness of the jaw, depending on whether the attachment is clamped on top of or under the jaw. The thickness of the jaw is marked on each instrument. Height gauges are available in a range of capacities reading from zero up to 1000 mm.

Vernier depth gauge

Accurate depths can be measured using the vernier depth gauge (Fig. 5.10) again employing the same principles. The fixed scale is similar to a narrow rule. The moving scale is tee-shaped to provide a substantial base and datum from which readings are taken. The instrument reading is the amount which the rule sticks out beyond the base. Depth gauges are available in a range of capacities from 150 mm to 300 mm.

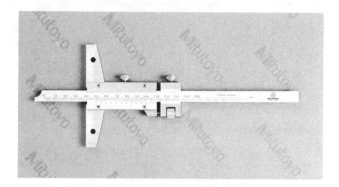

Figure 5.10 Vernier depth gauge

These are also available as an easy-to-read dial depth gauge (Fig. 5.11) and an electronic model with an LCD digital readout (Fig. 5.12) operating in the same way as the caliper models.

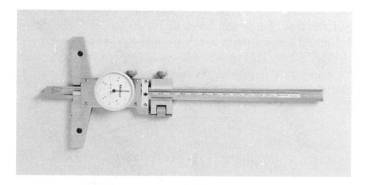

Figure 5.11 Dial depth gauge

Vernier bevel protractor

As well as linear measurement, vernier scales can equally well be used to determine angular measurement. The vernier bevel protractor (Fig. 5.13) again uses the

Figure 5.12 Electronic depth gauge

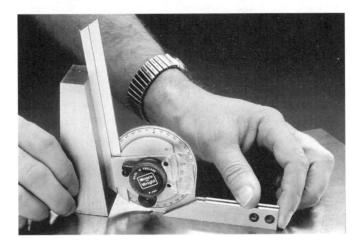

Figure 5.13 Vernier bevel protractor

principle of two scales, one moving and one fixed. The fixed scale is graduated in degrees, every 10 degrees being numbered 0, 10, 20, 30, etc. The moving or vernier scale is divided into 12 equal parts which occupy the same space as 23 degrees on the fixed scale (Fig. 5.14). This means that each division on the vernier scale is $\frac{23}{12}$ degrees $= 1\frac{11}{12}$ degrees or 1 degree 55 minutes. This is 5 minutes less than two divisions on the fixed scale (Fig. 5.15(a)).

Figure 5.14 Vernier protractor scale

If the two scales initially have their zeros in line and the vernier scale is then moved so that its first graduation lines up with the 2 degree graduation on the fixed

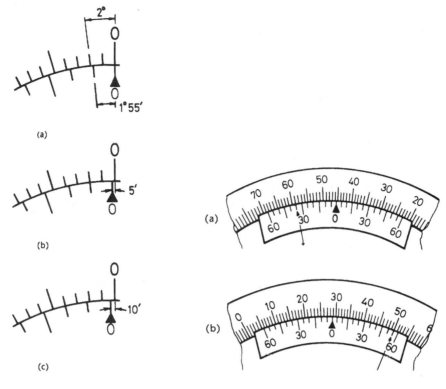

Figure 5.15 Vernier protractor scale readings

Figure 5.16 Vernier protractor readings

scale, the zero on the vernier scale will have moved 5 minutes (Fig. 5.15(b)). Likewise, the second graduation of the vernier lined up with the 4 degree graduation will result in the vernier scale zero moving 10 minutes (Fig. 5.15(c)) and so on until when the twelfth graduation lines up the zero will have moved $12 \times 5 = 60$ minutes = 1 degree. Since each division on the vernier scale represents 5 minutes, the sixth graduation is numbered to represent 30 minutes and the twelfth to represent 60 minutes.

The stock of the vernier protractor carries the fixed scale. The removable blade is attached to the moving or vernier scale, which has a central screw to lock the scale at any desired position and give angular measurement to an accuracy of 5 minutes.

Since the vernier scale can be rotated in both directions, the fixed scale is graduated from 0–90, 90–0, 0–90, 90–0 through 360°. This requires a vernier scale for each, and therefore the vernier scale is also numbered 0–60 in each direction.

To take a reading, note how many degrees the zero on the vernier scale is from the zero on the fixed scale. Then counting in the same direction, note the number of divisions on the vernier scale from zero to a line which exactly coincides with a line on the fixed scale.

In the reading shown in Fig. 5.16(a) the vernier scale has moved to the left 45 degrees. Counting along the vernier scale in the same direction, i.e. to the left, the seventh line coincides with a line on the fixed scale. Thus $7 \times 5 = 35$ minutes is added, to give a total reading of 45 degrees 35 minutes.

In the reading shown in Fig. 5.16(b), the vernier scale has moved to the right 28 degrees. Again counting along the vernier scale in the same direction, i.e. to the right, the eleventh line coincides with a line on the fixed scale, giving a total reading of 28 degrees 55 minutes.

5.2 Micrometers

The micrometer relies for its measuring accuracy on the accuracy of the spindle screw thread. The spindle is rotated in a fixed nut by means of the thimble, which opens and closes the distance between the ends of the spindle and anvil (see Fig. 5.17). The pitch of the spindle thread, i.e. the distance between two consecutive thread forms, is 0.5 mm. This means that, for one revolution, the spindle and the thimble attached to it will move a longitudinal distance of 0.5 mm.

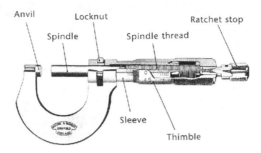

Figure 5.17 External micrometer

On a 0–25 mm micrometer, the sleeve around which the thimble rotates has a longitudinal line graduated in mm from 0 to 25 mm on one side of the line and subdivided in 0.5 mm intervals on the other side of the line.

The edge of the thimble is graduated in 50 divisions numbered 0, 5, 10, up to 45, then 0. Since one revolution of the thimble advances the spindle 0.5 mm, one graduation on the thimble must equal $0.5 \div 50$ mm $= 0.01$ mm. A reading is therefore the number of 1 mm and 0.5 mm divisions on the sleeve uncovered by the thimble plus the hundredths of a millimetre indicated by the line on the thimble coinciding with the longitudinal line on the sleeve.

In the reading shown in Fig. 5.18(a), the thimble has uncovered 9 mm on the sleeve. The thimble graduation lined up with the longitudinal line on the sleeve is $44 = 44 \times 0.01 = 0.44$. The total reading is therefore 9.44 mm.

Similarly in Fig. 5.18(b) the thimble has uncovered 16 mm and 0.5 mm and the thimble is lined up with graduation $27 = 27 \times 0.01 = 0.27$ mm, giving a total reading of 16.77 mm.

Greater accuracy can be obtained with external micrometers by providing a vernier scale on the sleeve. The vernier consists of five divisions on the sleeve, numbered 0, 2, 4, 6, 8, 0, these occupying the same space as nine divisions on the thimble, (Fig. 5.19(a)). Each division on the vernier is therefore equal to $0.09 \div 5 = 0.018$ mm. This is 0.002 mm less than two divisions on the thimble.

To take a reading from a vernier micrometer, note the number of 1 mm and 0.5 mm divisions uncovered on the sleeve and the hundredths of a millimetre on the thimble as with an ordinary micrometer. You may find that the graduation on the thimble does not exactly coincide with the longitudinal line on the sleeve, and this difference is obtained from the vernier. Look at the vernier and see which graduation

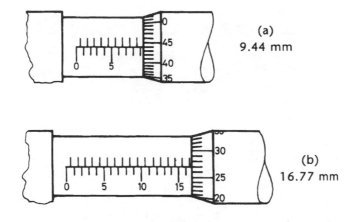

(a)
9.44 mm

(b)
16.77 mm

Figure 5.18 Micrometer readings

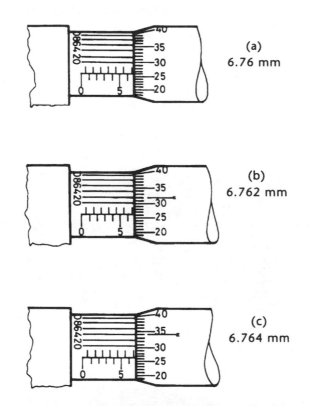

(a)
6.76 mm

(b)
6.762 mm

(c)
6.764 mm

Figure 5.19 Vernier micrometer readings

coincides with a graduation on the thimble. If it is the graduation marked 2, then add 0.002 mm to your reading (Fig. 5.19(b)); if it is the graduation marked 4, then add 0.004 mm (Fig. 5.19(c)); and so on.

External micrometers with fixed anvils are available with capacities ranging from 0–13 mm to 575–600 mm. External micrometers with interchangeable anvils (Fig. 5.20) provide an extended range from two to six times greater than the fixed-anvil types. The smallest capacity is 0–50 mm and the largest 900–1000 mm. To ensure

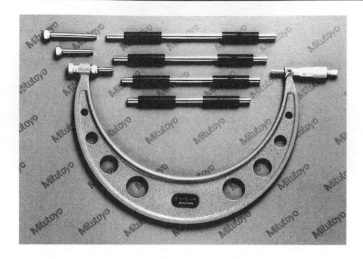

Figure 5.20 External micrometer with interchangeable anvils

accurate setting of the interchangeable anvils, setting gauges are supplied with each instrument.

Micrometers are available which give a direct mechanical readout on a counter (Fig. 5.21). The smallest model of 0–25 mm has a resolution of 0.001 mm and the largest of 125–150 mm a resolution of 0.01 mm.

Modern-designed external micrometers (Fig. 5.22) are available with an LCD digital readout giving direct readings in either imperial or metric units with a resolution of 0.00005" or 0.001 mm.

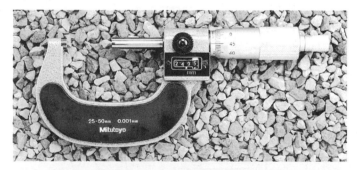

Figure 5.21 Direct reading digital micrometer

Figure 5.22 Electronic external micrometer

Internal micrometer

The internal micrometer (Fig. 5.23) is designed for inside measurement and consists of a micrometer measuring head to which may be added external rods to cover a wide range of measurements and a spacing collar to make up for the limited range of the micrometer head. Micrometers of less than 300 mm are supplied with a handle to reach into deep holes. Each extension rod is marked with the respective capacity of the micrometer when that particular rod is used. The smallest size is 25–50 mm with a measuring range of 7 mm. The next size covers 50–200 mm with a measuring range of 13 mm, while the largest covers 200–1000 mm with a measuring range of 25 mm.

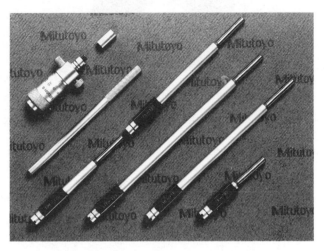

Figure 5.23 Internal micrometer

Readings are taken in the same way as described for the external micrometer, although, as already stated, the measuring range of the micrometer head is reduced.

Great care must be taken when using this instrument, a each of the measuring anvils has a spherical end, resulting in point contact. Experience in use is essential to develop a 'feel', and the instrument must be moved slightly back and forth and up and down to ensure that the measurement is taken across the widest point.

Depth micrometer

The depth micrometer (Fig. 5.24) is used for measuring the depths of holes, slots, recesses, and similar applications. Two types are available: one with a fixed spindle and a capacity of 0–25 mm, the other with interchangeable rods giving a measuring capacity up to 300 mm. The interchangeable rods are fitted into the instrument by unscrewing the top part of the thimble and sliding the rod in place, ensuring that the top face of the thimble and the underside of the rod are perfectly clean. The top of the thimble is then replaced and holds the rod in place. Each rod is marked with its respective size.

The micrometer principle is the same as for the other instruments; however, the readings with this instrument increase as the thimble is screwed on, resulting in the numbering of sleeve and thimble graduations in the opposite direction to those on the external and internal micrometers. To take a reading, you must note the 1 mm and 0.5 mm divisions covered by the thimble and add to this the hundredths of a

Figure 5.24 Depth micrometer

millimetre indicated by the line on the thimble coinciding with the longitudinal line on the sleeve.

In the reading shown in Fig 5.25(a) the thimble has covered up 13 mm and not quite reached 13.5. The line on the thimble coinciding with the longitudinal line on the sleeve is 44, so $44 \times 0.01 = 0.44$ mm is added, giving a total reading of 13.44 mm. Similarly, in Fig. 5.25(b) the thimble has just covered 17 mm and line 3 on the thimble coincides, giving a total reading of 17.03 mm.

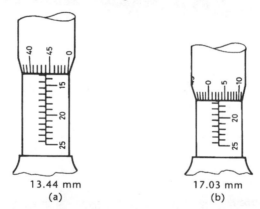

13.44 mm
(a)

17.03 mm
(b)

Figure 5.25 Depth micrometer readings

Depth micrometers are available giving a direct mechanical readout on a counter with a resolution of 0.01 mm (see Fig. 5.26).

Depth micrometers are also available as an electronic model with an LCD digital readout (Fig. 5.27) giving direct readings in imperial or metric units with a resolution of 0.0001" or 0.001 mm.

5.3 Dial indicators

Dial indicators magnify small movements of a plunger or lever and show this magnified movement by means of a pointer on a graduated dial. This direct reading from the pointer and graduated dial gives the operator a quick, complete, and

Figure 5.26 Direct reading digital depth micrometer

Figure 5.27 Electronic depth micrometer

accurate picture of the condition of the item under test. Dial indicators are used to check the dimensional accuracy of workpieces in conjunction with other equipment such as gauge blocks, to check straightness and alignments of machines and equipment, to set workpieces in machines to ensure parallelism and concentricity and for a host of other uses too numerous to list completely.

The mechanism of a dial indicator is similar to that of a watch and, although made for workshop use, care should be taken to avoid dropping or knocking it in any way. Slight damage to the mechanism can lead to sticking which may result in incorrect or inconsistent readings.

Plunger-type instruments

The most common instrument of this type is shown in Fig. 5.28. The vertical plunger carries a rack which operates a system of gears for magnification to the pointer. The dial is attached to the outer rim, known as the bezel, and can be rotated so that zero

Figure 5.28 Plunger-type dial indicator

Figure 5.29 Dial test indicator and stand

Figure 5.30 Back plunger-type dial indicator

Figure 5.31 Electronic-type indicator

can be set irrespective of the initial pointer position. A clamp is also supplied to prevent the bezel moving once it has been set to zero. The dial divisions are usually 0.01 or 0.002 mm, with an operating range between 8 and 20 mm, although instruments with greater ranges are available.

In conjunction with a robust stand or surface gauge (Fig. 5.29) this instrument can be used to check straightness, concentricity, as well as workpiece heights and roundness.

It may not always be possible to have the dial of this type facing the operator, which may create problems in reading the instrument or safety problems if the operator has to bend over equipment or a machine. An instrument which can be used to overcome these difficulties is the back plunger-type shown in Fig. 5.30. The readings can be seen easily by viewing above the instrument. The direction of the plunger movement restricts the range to about 3 mm.

Modern-design electronic plunger-type instruments (Fig. 5.31) are available with an LCD digital readout giving direct readings in imperial or metric units with a resolution of 0.0005" or 0.01 mm. These instruments can be zeroed at any point within the range.

Lever-type instruments

The lever-type of instrument is shown in Fig. 5.32. Due to the leverage system, the range of this type is not as great as that of the plunger-type and is usually 0.5 or 0.8 mm. The dial divisions are 0.01 or 0.005 mm, and again the dial is adjustable to set zero. The greatest advantage of this type is the small space within which it can work. Another added advantage is an automatic reversal system which results in movement above or below the contact stylus registering on the dial pointer. This facility, together with the contact stylus being able to swing at an angle, means that checks can be made under a step as well as on top (Fig. 5.33).

Figure 5.32 Lever-type dial indicator

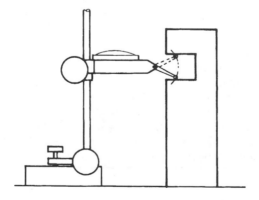

Figure 5.33 Lever-type dial indicator application

Table 5.1 shows a summary of measuring instruments.

Table 5.1

Instrument	Advantages	Limitations
Vernier calipers	Large measuring range on one instrument. Capable of internal, external, step and depth measurements. Resolution of LCD model 0.01 mm	Resolution 0.02 mm. Point of measuring contact not in line with adjusting nut. Jaws can spring. Lack of 'feel'. Length of jaws limits measurement to short distance from end of workpiece. No adjustment for wear
Vernier height gauge	Large range on one instrument. Resolution of LCD model 0.01 mm	Resolution 0.02 mm. No adjustment for wear
Vernier depth gauge	Large range on one instrument. Resolution of LCD model 0.01 mm	Resolution 0.02 mm. Lack of 'feel'. No adjustment for wear
Bevel protractor	Accuracy 5 minutes over range of 360°. Will measure internal and external angles	Can be difficult to read the small scales except with the aid of a magnifying lens
External micrometer	Resolution 0.01 mm or, with vernier 0.002 mm or LCD model 0.001 mm. Adjustable for wear. Ratchet or friction thimble available to aid constant 'feel'	Micrometer head limited to 25 mm range. Separate instruments required in steps of 25 mm or by using interchangeable anvils
Internal micrometer	Resolution 0.01 mm. Adjustable for wear. Can be used at various points along length of bore	Micrometer head on small sizes limited to 7 and 13 mm range. Extension rods and spacing collar required to extend capacity. Difficult to obtain 'feel'
Depth micrometer	Resolution 0.01 mm or with LCD model 0.001 mm. Adjustable for wear. Ratchet or friction thimble available to aid constant 'feel'	Micrometer head limited to 25 mm range. Interchangeable rods required to extend capacity
Dial indicator	Resolution as high as 0.001 mm. Measuring range up to 80 mm with plunger types. Mechanism ensures constant 'feel'. Easy to read. Quick in use if only comparison is required	Has to be used with gauge blocks to determine measurement. Easily damaged if mishandled. Must be rigidly supported in use.

6 Cutting tools and cutting fluids

The study of metal cutting is complex, due to the number of possible variables. Differences in workpiece materials and cutting-tool materials, whether or not a cutting fluid is used, the relative speed of the work and cutting tool, the depth of cut, and the condition of the machine all affect the cutting operation. However, certain basic rules apply, and when you know and can apply these you will be in a better position to carry out machining operations effectively.

The material from which the workpiece is made is not usually your choice. The operation to be carried out decides which machine you will use. This narrows the problem – knowing the operation and the machine, you can select the type of cutting tool; knowing the workpiece material you can decide the cutting-tool material, the cutting angles, the speeds at which to run the workpiece or cutting tool, and whether to use a cutting fluid. Finally, you must be able to maintain the cutting tools in good condition as the need arises, and this requires a knowledge of regrinding the tool usually by hand, known as off-hand grinding.

6.1 Cutting-tool materials

To be effective, the material from which a cutting tool is made must possess certain properties, the most important of which are red hardness, abrasion resistance, and toughness.

Red hardness: It is obvious that a cutting tool must be harder than the material being cut, otherwise it will not cut. It is equally important that the cutting tool remains hard even when cutting at high temperatures. The ability of a cutting tool to retain its hardness at high cutting temperatures is known as red hardness.

Abrasion resistance: When cutting, the edge of a cutting tool operates under intense pressure and will wear due to abrasion by the material being cut. Basically, the harder the material the better its resistance to abrasion.

Toughness: A cutting-tool material which is extremely hard is unfortunately also brittle. This means that a cutting edge will chip on impact if, for example, the component being machined has a series of slots and the cut is therefore intermittent. To prevent the cutting edge from chipping under such conditions, it is necessary that the material has a certain amount of toughness. This can be achieved only at the expense of hardness; that is, as the toughness is increased so the hardness decreases.

It can be readily seen that no one cutting tool material will satisfy all conditions at one time. A cutting tool required to be tough due to cutting conditions will not be at its maximum hardness and therefore not be capable of fully resisting abrasion. Alternatively, a cutting tool requiring maximum hardness will have maximum abrasion resistance but will not be tough to resist impact loads. The choice of cutting-tool material is governed by the type of material to be cut and the conditions under which cutting is to take place, as well as the cost of the tool itself. Remember that cutting tools are expensive, and great care should be taken to avoid damage and consequent wastage both in use and during resharpening.

High-speed steels (HSS)

High-speed tool steels consist of iron and carbon with differing amounts of alloying elements such as tungsten, chromium, vanadium, and cobalt. When hardened, these steels are brittle and the cutting edge will chip on impact or with rough handling. They have a high resistance to abrasion but are not tough enough to withstand high shock loads. These steels will cut at high speeds and will retain their hardness even when the cutting edge is operating at temperatures around 600°C.

A general-purpose, high-speed tool steel used to manufacture drills, reamers, taps, milling cutters, and similar cutting tools contains 18% tungsten, 4% chromium, and 1% vanadium and is referred to as an 18–4–1 tool steel. The addition of 8% cobalt to the above high-speed steel increases the hardness and red hardness and produces a steel which can be used at higher speeds. These tool steels are often referred to as super-high-speed steels.

Apart from being used to manufacture the cutting tools already mentioned, high-speed steel is available as 'tool bits' in round or square section already hardened and tempered. The operator has only to grind the required shape on the end before using.

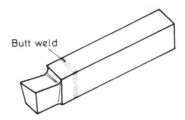

Butt weld

Figure 6.1 Butt-welded lathe tool

To save on cost, cutting tools such as lathe tools are made in two parts, instead of from a solid piece of expensive high-speed steel. The cutting edge at the front is high-speed steel and this is butt-welded to a tough steel shank. These tools are known as butt-welded tools, Fig. 6.1. The cutting edge can be reground until the high-speed steel is completely used and the toughened shank is reached. At this stage the tool is thrown away.

Stellite

Stellite is a cobalt–chromium–tungsten alloy containing no iron. It cannot be rolled or forged and is shaped by casting, from which it derives its cutting properties and hardness. No other form of heat treatment is required. Stellite is as hard as high-speed steel and has a higher red hardness, retaining its hardness at temperatures of 700°C. Being cast, it is also brittle and care must be taken to avoid chipping the cutting edge. It is more expensive than high-speed steel and is supplied as solid cast tool bits of round, square, and rectangular section, as tips, and as tipped tools where the tip is brazed to a toughened steel shank.

Stellite can be reground using standard grinding wheels, but care must be taken to avoid overheating, which leads to surface cracking and subsequent breakdown of the cutting edge.

Cemented carbides

Cemented carbides are produced by a powder-metallurgy technique, i.e. by using metals in their powder form. The final mixture of powders consists of various amounts of hard particles and a binding metal. The hard particles give the material its hardness and abrasion resistance, while the binding metal provides the toughness.

The most common hard particle used is tungsten carbide, but titanium carbide and tantalum carbide are often added in varying amounts. The binding metal used is cobalt, and various grades of cemented carbide are obtained for cutting different groups of materials by mixing in different proportions.

Cemented carbides normally contain 70–90% of hard particles together with 10–30% cobalt binding metal. In general, the more cobalt that is present, the tougher the cemented carbide. Unfortunately, however, this increase in toughness, obtained by increasing the cobalt content, results in decreased hardness and abrasion resistance.

Cemented carbides are used in cutting tools for turning, milling, drilling, boring, etc. in the form of tips or inserts which are brazed or clamped to a suitable tool shank, Fig. 6.2. The blanks are produced by mixing the metal powders in the correct proportions, pressing them into the required shape, and finally heating at temperatures as high as 1600°C, a process known as sintering. This sintering stage results in the cobalt binding metal melting and fusing with the hard particles, or cementing, to form a solid mass – hence the name cemented carbides.

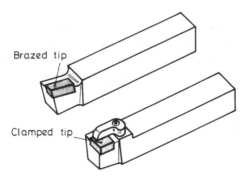

Brazed tip

Clamped tip

Figure 6.2 Tipped lathe tools

Cemented carbides are classified into three main groups: those used for machining steel, designated by the letter P and coloured blue; those used for machining cast iron and non-ferrous metals, designated K and coloured red; and finally multi-purpose grades, designated M and coloured yellow. These letters are followed by a number which, as it increases, denotes increasing toughness with a resultant decrease in hardness.

Cemented carbides have a red hardness higher than both high-speed steel and stellite and will retain their hardness at temperatures well in excess of 700°C.

Cemented-carbide tips are available in which a thin layer of titanium nitride is bonded over all the surfaces. This coating is extremely hard and has a low coefficient of friction, leading to an increase in abrasion resistance and a longer lasting cutting edge. It is claimed that speeds can be increased by 50% above those used with conventional cemented carbides with the cutting edge lasting the same time, or, alternatively, the cutting edge will last twice as long as that of the conventional tip if run at the same speed.

Owing to their extreme hardness, cemented carbides cannot be reground using the same wheels used to regrind high-speed steel and stellite. Silicon-carbide wheels (usually green in colour) must be used – these maintain a keen cutting action. To finish the cutting edge it is necessary to use a diamond wheel which laps the surfaces to produce a keen edge. Great care must be taken to avoid overheating, which leads to surface cracking of the tip and subsequent breakdown of the cutting edge.

Brazed-tip tools are expensive and have the disadvantages that they have to be removed from the machine to be reground and must then be reset in the machine, and after each regrind the tip becomes smaller and smaller. Tips which are clamped in a suitable holder – known as throw-away tips or inserts – do not suffer from these disadvantages. When one cutting edge is worn, the insert is merely unclamped, turned to the next keen cutting edge, and reclamped. This is repeated until all cutting edges are used (there may be as many as eight) and the insert is then thrown away. The holder is not removed from the machine, and no resetting is necessary. Both types do, however, have their applications in industry.

Ceramic cutting materials

Two types of ceramic cutting materials are available: a material made from pure aluminium oxide and a mixed ceramic of titanium carbide and aluminium oxide. The latter is used to cut the higher-strength steels and chilled cast iron. The ceramic powders are mixed, pressed into the required shape, and finally sintered, resulting in a solid dense blank which is subsequently ground to the correct size.

The tip blanks are used by clamping to a suitable tool holder.

Ceramic cutting materials have a high abrasion resistance and high red hardness. They show no deformation even at temperatures up to 1000°C, remaining hard at temperatures which would affect cemented carbides. They can be used to cut grey cast iron, spheroidal-graphite cast iron, malleable iron, and alloy steels at cutting speeds from 100 to 600 m/min at cutting depths of up to 6 mm in cast iron.

Use of a cutting fluid is not recommended because of the danger of thermal shock – pure aluminium oxide will be destroyed by a sudden temperature change of more than 200°C.

Cubic boron nitride

Next to diamond, cubic boron nitride is the hardest known material, with exceptionally high abrasion resistance and cutting-edge life in severe cutting conditions.

A layer of cubic boron nitride approximately 0.5 mm thick is bonded to a cemented-carbide tip approximately 5 mm thick. The cemented carbide provides a shock-resisting base. This material will machine chilled cast iron and fully hardened steel and still maintain a cutting edge. It is designed to perform most effectively on materials difficult to cut, and in some cases can replace a grinding operation.

It does not react with other metals or oxidise at temperatures below 1000°C and it is therefore virtually unaffected by heat generated in the high-speed cutting of difficult-to-machine materials.

Diamond

Diamond is the hardest known material. For this reason, single-crystal natural-diamond tools have been used in industry for a great number of years, to dress grinding wheels and as cutting tools to finish-machine non-ferrous and non-metallic materials.

A synthetic or man-made diamond material is now available which is extremely tough with a hardness approaching that of natural diamond. A layer of this synthetic diamond material is bonded to a tough shock-resisting base of cemented carbide for use in the form of tips. The range of application is on non-ferrous metals such as aluminium alloys, magnesium alloys, copper, brass, bronze, and zinc alloys and non-metallic materials such as ceramics, porcelain, and plastics. This material will also machine fully sintered tungsten carbide.

High cutting speeds are employed – up to 1200 m/min on non-ferrous metals, while sintered tungsten carbide can be machined at between 150 and 500 m/min depending upon the conditions. Under the same cutting conditions on the materials listed above, this synthetic diamond material will outlast all other cutting-tool materials, including single-crystal natural diamond, which is prone to chipping at the cutting edge.

6.2 Cutting tools

Clearance

All cutting tools whether held by hand or in a machine, must possess certain angles in order to cut efficiently. The first essential is a clearance angle, which is the angle between the cutting edge and the surface of the material being cut. This prevents any part of the cutting tool other than the cutting edge from coming in contact with the work, and so eliminates rubbing.

If the end of a cutting tool is ground parallel to the workpiece as shown in Fig. 6.3(a), the tool will skid along the work surface. If the back of the tool, or heel, is ground below the level of the cutting edge, it will rub on the work surface as shown in Fig. 6.3(b). The correct angle is shown in Fig. 6.3(c), where the heel of the tool is above the level of the cutting edge, thus leaving only the cutting edge in contact with the work.

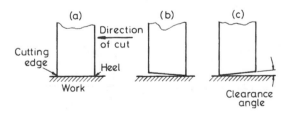

Figure 6.3 Clearance angles

The clearance angle should be kept at an absolute minimum, 8° being quite adequate for most purposes. Grinding an excessive clearance angle should be avoided – it is a waste of expensive cutting-tool material, a waste of time and money in grinding it in the first place, and, finally and most important, it weakens the cutting edge. In some cases, however, a greater clearance angle may be required, e.g. where holes are being machined using a boring tool, Fig. 6.4(a). If this additional clearance is provided up to the cutting edge, Fig. 6.4(b), serious weakening will result, so it is customary in these instances to provide the usual clearance angle for a short distance behind the cutting edge, known as primary clearance, followed by a second angle known as a secondary clearance angle, Fig. 6.4(c).

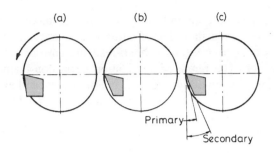

Figure 6.4 Primary and secondary clearance

Rake

For effective cutting, a second angle known as the rake angle is required. This is the angle between the tool face and a line at right angles to the surface of the material being cut.

The face upon which this angle is ground is the face along which the chip slides as it is being removed from the work. This angle therefore varies with the material being cut, since some materials slide more easily than others, while some break up into small pieces. Brass, for instance, breaks up into small pieces, and a rake angle of 0° is used. Aluminium, on the other hand, has a tendency to stick to the face of the tool and requires a steep rake angle, usually in the region of 30°.

For the majority of purposes, the rake angle used is positive, as shown in Fig. 6.5(a). When machining tough materials using the cemented-carbide cutting tools, it is necessary, due to the brittle nature of the carbide, to give maximum support to the tip. To achieve this, a negative rake is used so that the tip is supported under the cutting edge, Fig. 6.5(b).

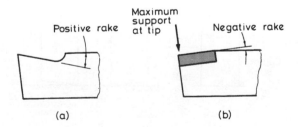

Figure 6.5 Positive and negative rake

Fig. 6.6 identifies the rake and clearance angles on various cutting tools.

The cutting angles of many cutting tools are established during their manufacture and cannot be changed by the user. Such tools include reamers, milling cutters, taps, and dies. These cutting tools can of course be resharpened, but a specialised tool-and-cutter grinding machine is required. The basic cutting tools used on centre

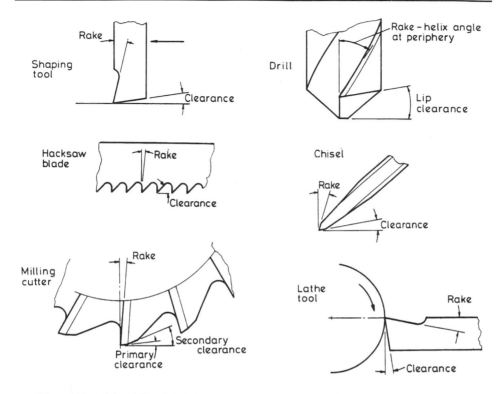

Figure 6.6 Identification of rake and clearance on various cutting tools

lathes and those used on the shaping machine are ground by hand, to give a variety of angles and shapes to suit different materials and applications.

Twist drills, although their helix angle is established during manufacture, are resharpened by grinding the point, the angle of which can be varied to suit different materials.

Turning and shaping tools

Cutting tools for use in turning may be required to cut in two directions. Such tools must therefore be provided with a rake and clearance angle for each direction of feed movement. This is illustrated in Fig. 6.7which shows a turning tool for facing and turning.

Facing will require the back rake and front clearance, since cutting takes place when the tool is feeding in the direction shown in Fig. 6.8(a). Turning will require side rake and side clearance, since cutting takes place when the tool is feeding in the direction shown in Fig. 6.8(b). Having the back and side rake on the same surface results in a true rake angle somewhere between the two, which is the angle along which the chip will flow when cutting in either direction. The trail angle is required to prevent the rear or trailing edge of the tool from dragging on the workpiece surface.

Cutting tools which are used to cut in only one feed direction require only one rake angle, although a number of clearance angles may be required to prevent rubbing. The knife tool shown in Fig. 6.9(a) acts in the direction shown, and a rake and a clearance angle are required in the same direction. Front clearance is also required to clear the workpiece surface.

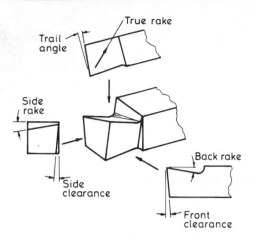

Figure 6.7 Turning tool clearance and rake angles

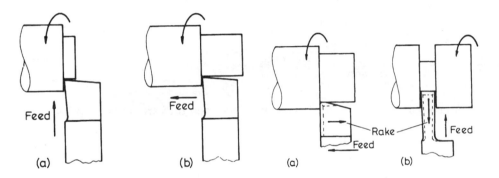

Figure 6.8 Direction of feed

Figure 6.9 Direction of rake angle on knife and undercut tools

A tool used to part-off or form undercuts requires rake and clearance in the direction of feed but also requires side clearance to prevent rubbing in the groove produced, Fig. 6.9(b).

A lathe tool is considered to be right-hand when it cuts from the right and left-hand when it cuts from the left, 6.10.

Cutting tools used in shaping are the same as those used in turning, except that the shanks are larger to withstand the shock at the start of each stroke. Cutting tools used across the face have rake angles as shown in Fig. 6.11(a), while those used to cut down a face are as shown in Fig. 6.11(b).

It is difficult to give precise values of rake angles, due to the number of variables encountered during machining. The values in Table 6.1 are offered as a guide for high-speed-steel cutting tools.

Twist drills

The nomenclature of the twist drill is shown in Fig. 6.12. The helix angle of the twist drill is the equivalent of the rake angle on other cutting tools and is established during manufacture. The standard helix angle is 30°, which, together with a point angle of 118°, is suitable for drilling steel and cast iron, Fig. 6.13(a).

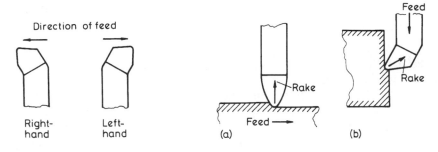

Figure 6.10 Right-hand and left-hand lathe tools

Figure 6.11 Rake angles of shaping tools

Table 6.1 Typical rake angles for high-speed-steel cutting tools

Material being cut	Rake angle
Brass	0°
Soft bronze	5°
Cast iron	8°
Mild steel	12°
Copper	20°
Aluminium	30°

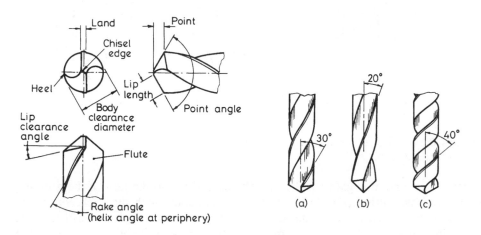

Figure 6.12 Nomenclature of twist drill

Figure 6.13 (a) Standard, (b) slow, and (c) quick helix drills

Drills with a helix angle of 20° – known as slow-helix drills – are available with a point angle of 118° for cutting brass and bronze, Fig. 6.13(b), and with a point angle of 90° for cutting plastics materials.

Quick-helix drills, with a helix angle of 40° and a point angle of 100°, are suitable for drilling the softer materials such as aluminium alloys and copper, Fig. 6.13(c).

Drill grinding

To produce holes quickly, accurately, and with a good surface finish, a correctly ground drill is required. In grinding the correct drill point, three important items must be controlled: the point angle, the lip length, and the lip clearance angle.

When a great deal of drilling is done, it is economical to use a drill-point grinding machine, which ensures correct point angle and lip clearance and equal lip lengths. However, it is often necessary to regrind a drill by hand, and you should be able to do this so that the drill cuts correctly.

The lip clearance angle is required to prevent rubbing. Too much will weaken the cutting edge; too little or none at all will result in rubbing and will produce excessive heat. In general, an angle of 10° to 12° gives the best results, but this can be increased to 15° for aluminium alloys and copper.

When grinding, it is important to ensure that the angle and length of each lip are equal. A drill having unequal angles or unequal lengths of lip, or both, results in an oversize hole. The effect of unequal lip lengths is shown in Fig. 6.14. As shown, the point is ground at the correct angle, but the unequal lengths of lip have the effect of placing the centre of the web off the centreline of the drill. The result is a hole which is oversize by an amount equal to twice the offset x, i.e. an offset as small as 0.25 mm results in a hole 0.5 mm oversize.

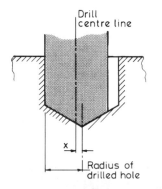

Figure 6.14 Effect of unequal lip lengths on a drill

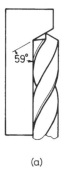 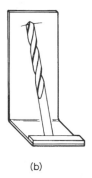

(a)   (b)

Figure 6.15 Angle and lip-length gauges

To produce a point angle of 118°, hold the drill at 59° to the face of the grinding wheel. Hold the cutting edge to be sharpened horizontal, with the back end of the drill lower than the front to create the correct lip clearance. Push the drill forward against the grinding wheel, and at the same time rock it slightly away from the cutting edge to give the lip clearance. Turn the drill round and repeat for the second lip. Repeat this a little at a time on each lip to get the correct angle and lip length.

A simple gauge as shown in Fig. 6.15(a) can be used as a guide to the correct point angle. Correct lip length can be checked using the simple gauge shown in Fig. 6.15(b). With the shank of the drill supported, a line is scratched with each lip. When the lines coincide, the lip lengths are equal.

6.3 Cutting-tool maintenance

An off-hand grinding machine is used with a workpiece or cutting tool held by hand and applied to the grinding wheel. Because of this, its use requires stringent safety precautions

The off-hand grinding machine is basically an electric motor having a spindle at each end, each carrying a grinding wheel, also referred to as an abrasive wheel. This arrangement allows a coarse wheel to be mounted at one end and a fine wheel at the other. All rough grinding is carried out using the coarse wheel, leaving the finishing operations to be done on the fine wheel.

These machines may be mounted on a bench, when they are often referred to as bench grinders, or on a floor-mounted pedestal and referred to as pedestal grinders.

The grinding wheel is adequately guarded to protect the operator in the event of the grinding wheel bursting, and the machine must *never* be run without these guards in position. Eye protection in the form of eye shields, goggles, or safety spectacles *must* be worn by any person operating an off-hand grinder.

An adjustable work rest is fitted to the front and sometimes to the sides the grinding wheel to support the cutting tool during grinding. The work rest is adjustable for angle, so that cutting tools rested on it can be ground at a specific angle. The work rest must be adjusted so that at all times the gap between it and the surfaces of the wheel is at a minimum, and it must be properly secured. This prevents the possibility of the workpiece, cutting tool, or fingers becoming jammed between the wheel and the work rest. The work rest must be adjusted frequently to compensate for wheel wear and maintain the gap at a minimum. All the above precautions are required by law under various regulations, and failure to comply with them can not only result in possible personal injury but may also lead to prosecution.

When using an off-hand grinder, use the outside diameter of the wheel wherever possible and keep the cutting tool moving across the wheel surface. Grinding in one place results in a groove being worn on the wheel surface, making it impossible to grind cutting edges straight. Use the side of the wheel only when you require a flat surface on the cutting tool.

Do not grind for a long period without cooling the cutting tool. Grind off small amounts and cool frequently to avoid overheating and the possibility of cracking.

The wheels fitted to off-hand grinders are usually chosen for general use to cover a range of materials. These general-grade wheels will not grind cemented carbides, for which a green-grit silicon-carbide wheel is necessary.

6.4 Cutting speed

The relative speed between the cutting tool and workpiece is known as the cutting speed and is expressed in metres per minute. In turning, the cutting speed is the surface speed of the work, i.e. the speed of a point on the circumference. In milling and drilling, it is the surface speed of the cutting tool.

As an example, we shall find the cutting speed S of a workpiece of diameter d mm being turned at N rev/min as shown in Fig. 6.16.

The distance travelled by a point on the circumference in one revolution = the circumference = πd mm.

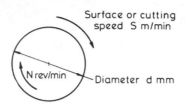

Figure 6.16 Cutting speed

In one minute the workpiece turns through N revs, so the distance travelled by the same point in one minute $= \pi dN$ mm/min.

But cutting speed S is expressed in metres per minute, so, dividing by 1000, we get

$$\text{cutting speed } S = \frac{\pi dN}{1000} \text{ m/min}$$

☐ *Example 6.1*
Find the cutting speed of a 50 mm diameter bar being turned with a spindle speed of 178 rev/min.

$$S = \frac{\pi dN}{1000} = \frac{\pi \times 50 \times 178}{1000} = 28 \text{ m/min}$$

☐ *Example 6.2*
Find the cutting speed of a 15 mm diameter drill running at 955 rev/min.

$$S = \frac{\pi dN}{1000} = \frac{\pi \times 15 \times 955}{1000} = 45 \text{ m/min}$$

The cutting speed is dictated mainly by the type of cutting-tool material being used and the type of material being machined, but it is also influenced by the depth of cut and the feed rate. Using high-speed-steel tools, aluminium can be machined at higher cutting speeds than steel, and at even higher cutting speeds using cemented-tungsten-carbide tools.

The manufacturers of cutting-tool materials give recommendations for the cutting speeds at which their tool materials will cut various work materials such as brass, aluminium, steel, and so on. A selection of these values if given for guidance in Table 6.2.

Table 6.2 Typical cutting speeds with HSS and tungsten-carbide tools

Material being cut	Cutting speed (m/min)	
	High-speed steel	*Tungsten carbide*
Cast iron	20	160
Mild steel	28	250
Bronze	35	180
Hard brass	45	230
Copper	60	330
Aluminium	100	500

Knowing the value of the cutting speed for a particular combination of cutting-tool material and work material, and knowing the diameter of the work to be produced (or the tool diameter), the unknown value is the spindle speed N at which the work or tool should be run. From the cutting-speed equation, this is given by

$$N = \frac{1000S}{\pi d}$$

Example 6.3

At what spindle speed would a 200 mm diameter high-speed-steel milling cutter be run to machine a steel workpiece, if the cutting speed is 28 m/min?

$$N = \frac{1000S}{\pi d} = \frac{1000 \times 28}{\pi \times 200} = 45 \text{ rev/min}$$

Example 6.4

What spindle speed would be required to turn a 150 mm diameter cast iron component using cemented-tungsten-carbide tooling at a cutting speed of 160 m/min?

$$N = \frac{1000S}{\pi d} = \frac{1000 \times 160}{\pi \times 150} = 340 \text{ rev/min}$$

You should adopt the habit of carrying out this calculation before starting any machining operation, even if only as a rough mental calculation. This way you at least have a basis for running the machine at a speed which is much nearer the correct speed than if you guess. When you gain experience, you will find that the spindle speed can be adjusted to suit different conditions by watching the way in which the chip is removed from the work.

6.5 Cutting fluids

During metal cutting, the metal immediately ahead of the cutting tool is severely compressed, which results in heat being generated. The metal then slides along the tool face, friction between the two surfaces generating additional heat. Any rubbing between the tool and the cut surface, which would occur with tool wear when the clearance angle is reduced, also produces heat. This heat is usually detrimental, especially to high-speed-steel cutting tools. Some metals, as they are cut, have a tendency to produce a chip which sticks or welds to the tool face, due chiefly to the high pressure between the metal and the tool. This has the effect of increasing the power required for cutting, increasing the friction and therefore heat, and finally, as the chip breaks away from the tool face and reforms, it creates wear on the tool face and a bad surface finish on the work. Excessive heat generated during the cutting may be sufficient to cause the work to expand. Work measured under these conditions may be undersize when it cools.

The basic role of a cutting fluid is to control heat, and it may do this by direct cooling of the work, chip, and tool or by reducing friction by lubricating between the work, chip, and tool. To cool effectively, a cutting fluid should have a high specific heat capacity and a high thermal conductivity.

The fluids most readily associated with cooling and lubricating are water and oil. Water has a higher specific heat capacity and thermal conductivity than oil, but unfortunately will promote rust and has no lubricating properties. Oil does not promote rust, has good lubricating properties, but does not cool as well as water. To benefit from the advantages of each, they can be mixed together with various additives to give a required measure of cooling and lubrication. With the high cost of oil, the cost of savings of water-based fluids are so great that a great deal of development is being carried out to provide fluids which have good lubricating properties when mixed with water.

In general, the use of cutting fluids can result in

- less wear on cutting tools,
- the use of higher cutting speeds and feeds,
- improved surface finish,
- reduced power consumption,
- improved control of dimensional accuracy.

The ideal cutting fluid, in achieving the above, should

- not corrode the work or machine,
- have a low evaporation rate,
- be stable and not foam of fume,
- not injure or irritate the operator.

6.6 Types of cutting fluid

Neat cutting oils

These oils are neat in so much as they are not mixed with water for the cutting operation. They are usually a blend of a number of different types of mineral oil, together with additives for extreme-pressure applications. Neat cutting oils are used where severe cutting conditions exist, usually when slow speeds and feeds are used or with extremely tough and difficult-to-machine steels. These conditions require lubrication beyond that which can be achieved with soluble oils.

In some cases soluble oil cannot be used, due to the risk of water mixing with the hydraulic fluid or the lubricating oil of the machine. A neat oil compatible with those of the machine hydraulic or lubricating system can be used without risk of contamination. Neat cutting oils do not have good cooling properties and it is therefore more difficult to maintain good dimensional accuracy. They are also responsible for dirty and hazardous work areas by seeping from the machine and dripping from workpieces and absorbing dust and grit from the atmosphere.

Low-viscosity or thin oils tend to smoke or fume during the cutting operation, and under some conditions are a fire risk.

The main advantages of neat cutting oils are their excellent lubricating property and good rust control. Some types do, however, stain non-ferrous metals.

Soluble oils

Water is the cheapest cooling medium, but it is unsuitable by itself, mainly because it rusts ferrous metals. In soluble oils, or more correctly emulsifiable oils, the excellent cooling property of water is combined with the lubricating and protective qualities of mineral oil. Oil is, of course, not soluble in water, but with the aid of an agent known as an emulsifier it can be broken down and dispersed as fine particles throughout the water to form an emulsion.

Other ingredients are mixed with the oil to give better protection against corrosion, resistance to foaming and attack by bacteria, and prevention of skin irritations. Under severe cutting conditions where cutting forces are high, extreme-pressure (EP) additives are incorporated which do not break down under these extreme conditions but prevent the chip welding to the tool face.

Emulsions must be correctly mixed, otherwise the result is a slimy mess. Having selected the correct ratio of oil to water, the required volume of water is measured into a clean tank or bucket and the appropriate measured volume of soluble oil is added gradually at the same time as the water is slowly agitated. This will result in a stable oil/water emulsion ready for immediate use.

At dilutions between 1 in 20 and 1 in 25 (i.e. 1 part oil in 20 parts water) the emulsion is milky white and is used as a general-purpose cutting fluid for capstan and centre lathes, drilling, milling, and sawing.

At dilutions from 1 in 60 to 1 in 80 the emulsion has a translucent appearance, rather than an opaque milky look, and is used for grinding operations.

For severe cutting operations, such as gear cutting or broaching and machining tough steels, fluids with EP additives are used at dilutions from 1 in 5 to 1 in 15.

As can readily be seen from the above, when the main requirement is direct cooling, as in the case of grinding, the dilution is greater, i.e. 1 in 80. When lubrication is the main requirement, as with gear cutting, the dilution is less i.e. 1 in 5.

The advantages of soluble oils over neat cutting oils are their greater cooling capacity, lower cost, reduced smoke, and elimination of fire hazard. Disadvantages of soluble oils compared with neat cutting oils are their poorer rust control and that the emulsion can separate, be affected by bacteria, and become rancid.

Synthetic fluids

Sometimes called chemical solutions, these fluids contain no oil but are a mixture of chemicals dissolved in water to give lubricating and anti-corrosion properties. They form a clear transparent solution with water, and are sometimes artificially coloured. They are very useful in grinding operations, where, being non-oily, they minimise clogging of the grinding wheel and are used at dilutions up to 1 in 80. As they are transparent, the operator can see the work, which is also important during grinding operations.

They are easily mixed with water and do not smoke during cutting. No slippery film is left on the work, machine, or floor. They give excellent rust control and do not go rancid. At dilutions of between 1 in 20 and 1 in 30 they can be used for general machining.

Semi-synthetic fluids

These are recently developed cutting fluids, sometimes referred to as chemical emulsions. Unlike synthetic fluids, these fluids do have a small amount of oil emulsified in water, as well as dissolved chemicals, but they are not true emulsions. When mixed with water they form extremely stable transparent fluids, with the oil in very small droplets. Like the synthetic types, they are often artificially coloured for easy recognition.

They have the advantage over soluble oil of increased control of rust and rancidity and a greater application range. They are safer to use, will not smoke, and leave no slippery film on work, machine, or floor. Depending upon the application, the dilution varies between 1 in 20 and 1 in 100.

6.7 Application of cutting fluids

Having selected the correct type of cutting fluid, it is equally important to apply it correctly. This is best done by providing a generous flow at low pressure to flood the work area. Flooding has the added advantage of washing away the chips produced. Fluid fed at high pressure is not recommended, since it breaks into a fine spray or mist and fails to cool or lubricate the cutting zone. To cope with the large flow of fluid, the machines must have adequate splash guards, otherwise the operator tends to reduce the flow and the resulting dribble does little to improve cutting.

Many methods have been used to direct the fluid into the cutting zone and from every possible direction. The shape of the nozzle is important but depends largely on the operation being carried out and on the shape of the workpiece. The nozzle may be a simple large-bore pipe or be flattened as a fan shape to provide a longer stream. The main flow may be split into a number of streams directed in different directions – up, down, or from the sides – or, by means of holes drilled in a length of pipe, create a cascade effect. In some cases, especially with grinding, where the wheel speed creates air currents which deflect the cutting fluid, deflector plates are fitted to the pipe outlet. Where the cutting tool is vertical, it can be surrounded by a pipe having a series of holes drilled into the bore and directed towards the cutting tool. Whatever the method used, the fundamental need is to deliver continuously an adequate amount of cutting fluid where it is required.

6.8 Safety in the use of cutting fluids

Cutting fluids can affect the health of those exposed to them in various ways: by contact with the skin, by contact with the eyes, if they are breathed in with air as small droplets or vapour, or if they are swallowed. The possible effects of these can normally be avoided by good housekeeping and a high standard of personal hygiene.

The following precautions, if observed, will reduce or eliminate the likely hazards.

- Working methods should be employed that avoid direct skin contact with oils, e.g. machine splash guards and correct handling procedures should be used.
- Adequate local exhaust ventilation should be provided for areas where vapour and mists are generated.
- Adequate protective clothing should be worn.
- Only disposable 'wipes' or clean rags should be used. Contaminated rags and tools should never be put into overall pockets.
- A barrier cream should be applied to the hands and exposed areas of the arms before starting work and on resuming work after a break.
- Hands should be thoroughly washed using suitable hand cleaners and warm water and dried using a clean towel before, as well as after, going to the toilet, before eating, and at the end of each shift.

- Conditioning cream, applied after washing, replaces fatty matter in the skin and helps prevent dryness.
- Contaminated clothing, especially undergarments, should be changed regularly and be thoroughly cleaned before re-use.
- Overalls should be cleaned frequently.
- Paraffin, petrol, and similar solvents should not be used for skin-cleansing purposes.
- All cuts and scratches must receive prompt medical attention.
- Your should seek prompt medical advice if you notice any skin abnormality.

7 | **Drilling**

The majority of drilling work is carried out on pillar drilling machines, so called because the machine elements are arranged on a vertical pillar. The machines in the heavy-duty range have power feed, are driven from the motor through a gearbox, and have a drilling capacity in steel up to 50 mm diameter. Smaller sensitive machines, Fig. 7.1 have a hand feed, giving the sensitivity, are belt driven from the motor through pulleys, and have a maximum drilling capacity in steel ranging from 5 mm up to 25 mm diameter. These machines may be bench- or floor-mounted.

7.1 The sensitive drilling machine

The main elements of a typical sensitive drilling machine are shown in Fig. 7.1.

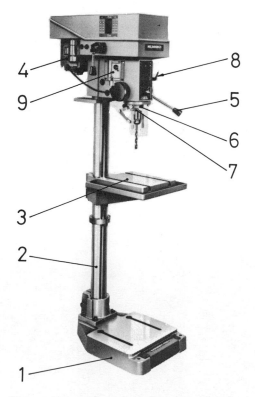

Figure 7.1 Sensitive drilling machine

1. Base – provides a solid foundation for the machine, into which the pillar is securely clamped.
2. Pillar – provides a solid support for the drill head and worktable.

3. Worktable – provides a flat surface in correct alignment with the drill spindle upon which the workpiece can be positioned. Tee slots are provided for clamping purposes. The worktable can be raised, lowered, and swung about the pillar and be securely clamped in the required position.

4. Motor – provides the drive to the spindle through a five-step pulley system and a two-speed gearbox, Fig. 7.2. Thus five pulley speeds with A and B in mesh and five with C and D in mesh give a range of ten spindle speeds from 80 to 4000 rev/min.

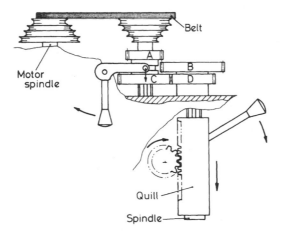

Figure 7.2 Drilling-machine drive system

5. Handwheel – provides feed to the drill by means of a rack and pinion on the quill, Fig. 7.2.

6. Quill – this is the housing inside which the spindle rotates. Only the longitudinal movement is transmitted by the quill, which itself does not rotate.

7. Spindle – provides the means of locating, holding, and driving the cutting tools and obtains its drive through the pulley.

8. Depth stop – provides a means of drilling a number of holes to a constant depth.

9. Stop/start – the machine shown is switched on by a shrouded push-button starter with a cover plate which can be padlocked to prevent unauthorised access. A mushroom-headed stop button is situated on the starter, and the machine can also be switched off using the emergency kick-stop switch at the front of the base. A safety switch is also incorporated under the belt guard and automatically stops the spindle should the guard be lifted while the machine is running.

7.2 Tool holding

Drills and similar tools with parallel shanks are held in a drill chuck, Fig. 7.3. Many different types of chuck are available, each being adjustable over its complete range, and give good gripping power. By rotating the outer sleeve, the jaws can be opened and closed. To ensure maximum grip, the chuck should be tightened using the correct size of chuck key. This prevents the drill from spinning during use and chewing up the drill shank.

A hazard in the use of chucks is the possibility of leaving a chuck key in position. When the machine is then switched on, the chuck key can fly in any direction and

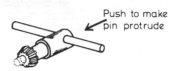

Figure 7.3 Drill chuck   **Figure 7.4** Safety chuck key

cause serious injury. When you remove a drill from the chuck, always remember to remove the chuck key. Never leave it in the chuck for even the shortest time. Better still, use a safety chuck key, Fig. 7.4, in which the central pin is spring-loaded and has to be pushed to engage. When the force is released, the pin retracts and the key falls from the chuck.

The chuck is fitted with a Morse-taper shank which fits into a corresponding Morse taper in the spindle. The size of Morse taper is identified from smallest to largest by numbers 1, 2, 3, 4, 5, and 6. The included angle of each taper is different but is very small, being in the region of 3°. If the two mating tapered surfaces are clean and in good condition, this shallow taper is sufficient to provide a drive between the two surfaces. At the end of the taper shank, two flats are machined, leaving a portion known as the tang. This tang fits in a slot on the inside of the spindle and its main purpose is for the removal of the shank.

To remove a shank from the spindle, a taper key known as drift is used. The drift is inserted through a slot in the spindle as shown in Fig. 7.5.

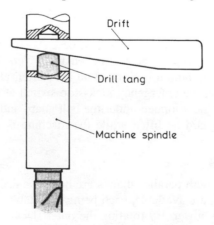

Figure 7.5 Drift in drill spindle

Drills are available with Morse-taper shanks which fit directly into the spindle without the need for a chuck. The size of Morse taper depends on the drill diameter, and the range is shown in Table 7.1.

Table 7.1 Range of Morse tapers

Morse taper	No. 1	No. 2	No. 3	No. 4	No. 5	No. 6
Drill-diameter range (mm)	up to	14.25 to	23.25 to	32 to	51 to	77 to
	14	23	31.75	50.5	76	100

It is essential that tapers are kept clean and in good condition. As already stated, the drive is by friction through the tapered surfaces, and any damage to these surfaces puts some of the driving force on the tang. If this force is excessive, the tang can be twisted off. When this happens the drill has to be discarded, as there is no way of easily removing it from the spindle.

Where a cutting tool or chuck has a Morse taper smaller than that of the spindle, the difference is made up by using a sleeve. For example, a drill with a No. 1 Morse-taper shank to be fitted in a spindle with a No. 2 Morse taper would require a 1–2 sleeve, i.e. No. 1 Morse-taper bore and a No. 2 Morse taper outside. Sleeves are available from 1–2, 1–3, 2–3, 2–4, and so on over the complete range.

7.3 Clamping

Work is held on a drilling machine by clamping to the worktable, in a vice or, in the case of production work, in a jig. It is sufficient to say here that work held in a jig will be accurately drilled more quickly than by the other methods, but large quantities of the workpiece must be required to justify the additional cost of the equipment.

Standard equipment in any workshop is a vice and a collection of clamps, studs, bolts, nuts, and packing. It should be stressed that work being drilled should never be held by hand. High forces are transmitted by a revolving drill, especially when the drill is breaking through the bottom surface, which can wrench the work from your hand. The resulting injuries can vary from a small cut to the loss of a finger.

Never take a chance – always clamp securely.

Small workpieces with parallel faces can be quite adequately held in a vice. The work is then positioned under the drill and the vice is clamped to the worktable.

Larger work and sheet metal are best clamped direct on to the worktable, care of course being taken to avoid drilling into the worktable surface. When required, the work can be raised off the worktable surface by means of suitable packing or on parallels. Tee slots are provided in the worktable surface into which are fitted tee bolts, or tee nuts in which studs are screwed, Fig. 7.6.

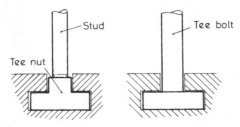

Figure 7.6 Tee nut and tee bolt

Various styles and shapes of clamp are available, one of which is shown in Fig. 7.7. The central slot enables it to be adjusted to suit the workpiece. To provide sound

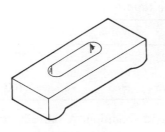

Figure 7.7 Clamp

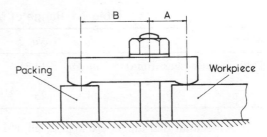

Figure 7.8 Clamping forces

clamping, the clamp should be reasonably level, and this is achieved by packing under the rear of the clamp to as near as possible the same height as the workpiece, Fig. 7.8.

The clamping bolt should be placed close to the work, since the forces on the work and packing are inversely proportional to their distances from the bolt. For greatest clamping force on the work, distance A in Fig. 7.8 must be less than distance B.

7.4 Cutting tools on drilling machines

Various cutting tools besides twist drills are used on a drilling machine, and some of them are described below.

Twist drill

Twist drills are available with parallel shanks up to 16 mm diameter and with taper shanks up to 100 mm diameter and are made from high-speed steel. Standard lengths are known as jobber-series twist drills, short drills are known as stub series, and long drills as long series and extra long series. Different helix angles are available for drilling a range of materials, as described in Chapter 6.

Combination drills known as Subland drills combine a number of operations in a single tool; for example drill and ream, drill two diameters, drill and chamfer, drill and spotface, drill and counterbore, Fig. 7.9. Each cutting edge has a separate land and flute, Fig. 7.10, which enables cutting to take place and resharpening to be easily carried out.

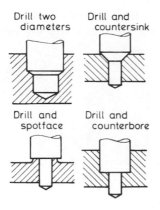

Figure 7.9 Examples of Subland-drill applications

Figure 7.10 Subland drill

Reamer

A reamer is used to produce a hole of greater accuracy than can be obtained using a drill. The hole is drilled undersize by an amount depending upon the diameter but usually about 0.4 mm; the required size is then obtained with the reamer. Care should be taken with the position and alignment of the drilled hole, since the reamer will correct size, roundness, and surface finish but will not correct errors in alignment.

As a general rule, reaming is carried out at half the speed used for drilling.

Reamers are made from high-speed steel in sizes up to 50 mm diameter in a variety of types, the most common being the machine reamer, Fig. 7.11. Machine reamers have a Morse-taper shank, although reamers below 12 mm diameter are available with parallel shanks. The diameter is constant along its length, and cutting takes place on the chamfer or bevel at the front, which is usually 45°. This chamfer can be reground using a cutter-grinding machine.

Figure 7.11 Machine reamer

Flutes on reamers usually have a left-hand spiral or helix, opposite to that of a drill. This pushes the metal chips ahead of the reamer rather than back up the flutes and prevents scratching the bore. This feature also prevents the reamer from 'screwing' itself into the hole, which would tend to happen with a right-hand helix.

Countersink

Countersink cutters, Fig. 7.12, made from high-speed steel, are used to cut a large chamfer of the correct angle, usually 90°, as a seating for countersink-head screws. Countersinks should be run at a fairly slow speed to avoid chatter. They are available with parallel and taper shanks.

Figure 7.12 Countersink cutter

Counterbore

A counterbore cutter, Fig. 7.13, is used to enlarge an existing hole to provide a flat and square seating for a screw, bolt, or nut under the workpiece surface. Teeth are provided on the end face and on the circumference, to permit cutting to a depth. A pilot is provided which locates in the existing hole and guides the tool during cutting. These pilots may be a solid part of the tool or detachable when the cutter is used on a series of different-size holes.

Counterbore cutters are made from high-speed steel and may have parallel or taper shanks.

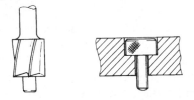

Figure 7.13 Counterbore cutter

Spotface

A spotface cutter, Fig. 7.14, is used to provide a flat and square seating for a screw, bolt, or nut on the surface of the workpiece, usually on the surface of a rough casting which would not otherwise provide a sufficiently accurate seating. The spotface is similar to a counterbore cutter, but has teeth on the end only. It will cut to only a very limited depth and cannot be used to counterbore. Counterbore cutters, on the other hand, can be used to spotface.

In some awkward places, back or underside spotfacing is required, Fig. 7.15. The pilot is fed through the hole and the cutter is fixed to the pilot, usually by some quick-locking mechanism or a simple grub screw. When the operation is finished, the cutter is removed and the pilot withdrawn.

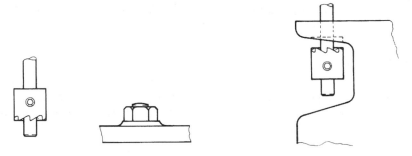

Figure 7.14 Spotface cutter   **Figure 7.15** Back spotfacing

Trepanning tools

Where large-diameter holes are required in sheet metal, they can be conveniently cut using a trepanning tool, Fig. 7.16(a).

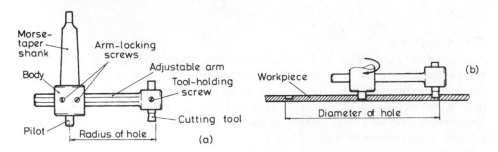

Figure 7.16 Trepanning tool

A small hole, to suit the pilot, is drilled at the centre of the required position. The adjustable arm is extended so that the edge of the high-speed-steel cutting edge produces the required size of hole. This may require a number of trials and adjustments before the correct size is reached. The pilot is located in the pilot hole and, as the tool rotates, it is fed through the work, Fig. 7.16(b).

The arm can be adjusted to any diameter within the range of the trepanning tool, which may be as large as 300 mm. The cutting tool can be ground to any angle to suit the material being cut.

Taps

Tapping can be carried out very efficiently on a drilling machine but requires the use of a special tapping attachment. The tapping attachment has a clutch which is preset according to the size of tap being used. When the tap hits the bottom of the hole or a hard piece of material, the clutch slips, causing the tap to remain stationary while the spindle keeps rotating. The machine is then switched to reverse and the tap is extracted. By using this attachment, broken taps are eliminated and the tapping operation is done more quickly and accurately than by hand.

7.5 Drilling operations

Unless the workpiece is held in a drill jig, the position of holes on a workpiece must be marked out. When the position of a hole is determined, its centre is indicated by means of a centre dot, using a centre punch. This centre dot is used to line up the drill and as a means of starting the drill in the correct position. The workpiece is set on the worktable, carefully positioned under the drill, using the centre dot, and clamped in position as shown in Fig. 7.17. Two clamps are usually required, one at each side of the component.

When held in a vice, the workpiece should be positioned and the vice be tightened securely. The workpiece is then positioned under the drill as before and the vice is clamped to the worktable.

In positioning the workpiece, take care to avoid drilling into the vice or worktable. If necessary, raise the workpiece on parallels, Fig. 7.17.

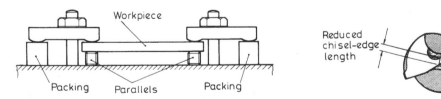

Figure 7.17 Component clamped on drilling machine

Figure 7.18 Drill-point thinning

Before drilling, check the drill is the correct size. Because you remove it from a space marked 5 mm does not mean it is a 5 mm drill – the person who used it before you may not have returned it to the correct space. Also, check the condition of the cutting edges and if necessary resharpen them.

Having carefully lined up and clamped the workpiece, begin drilling, taking care that the drill is still central with your required position. Small-diameter drills will start in the correct position with the aid of the centre dot; large-diameter drills with a long chisel edge require other means to assist in starting. The best method is to use a smaller-diameter drill on the centre dot, but stop before it cuts to its full diameter. The larger drill will start in its correct position guided by the 118° dimple produced by the smaller drill.

Where the chisel edge is found to be too wide for a particular purpose, it can be reduced by point thinning, Fig. 7.18. This can be done using the edge of a well-dressed grinding wheel, but it is perhaps better left to a more experienced person.

A properly sharpened drill run at the correct speed will produce a spiral type of chip from each cutting edge. As the hole becomes deeper, the chips tend to pack in the flutes and the drill may have to be removed from the hole periodically to clear the chips.

Most trouble in drilling arises when the drill breaks through the far surface. The chisel edge of a drill centres and guides it through the workpiece, keeping the hole straight. When the chisel edge breaks through, it can no longer guide the drill and keep it central, and the drill will wobble and bounce in its own hole, an occurrence known as 'chatter'.

When the complete drill point has almost broken through, there is a tendency for the drill to 'snatch' or 'grab'. This happens when the metal still to be cut is so thin that it is pushed aside rather than cut and the drill pulls itself through due to the helix angle – in the same way as a screw thread advances. In the case of unclamped work, it would be the workpiece which would be pulled up the drill, wrenching it out of the hand holding it, with resulting injury or breakage, or both.

A repeat warning here – always clamp the workpiece.

To avoid these problems when breaking through, take great care and avoid too rapid a feed.

Holes can be drilled to a particular depth by setting the depth stop on the machine. The workpiece is positioned as already described, and drilling is started until the drill is just cutting its full diameter. The machine is switched off and the stationary drill is brought down into contact with the workpiece. The depth stop is set to the required dimension by adjusting it to leave the required space above the spindlehead casting and it is then locked in position, Fig. 7.19.

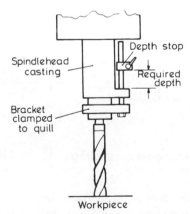

Figure 7.19 Depth stop

Where holes in two parts are required to line up with each other, a technique known as 'spotting' is carried out. The top part is marked out and drilled as already described. The two parts are then carefully positioned and clamped together. The holes in the bottom part are then transferred by 'spotting' through from the top part.

Drilling of the bottom part can then proceed in the knowledge that both sets of holes are identical, which may not be the case if both parts are marked out and drilled individually.

When the two parts are to be screwed together, the bottom part requires to be tapped while the top part requires a clearance hole. The sequence is the same as for spotting except that, having positioned, clamped, and spotted with the clearance drill, the drill is changed to the tapping size. The hole is then drilled and tapped, Fig. 7.20.

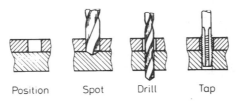

Position   Spot   Drill   Tap

Figure 7.20 Spotting through from existing workpiece

7.6 Drilling sheet metal

The same problems already discussed when the drill breaks through apply to drilling sheet metal. The problems are increased with thin sheet, since the chisel edge can break through before the drill is cutting its full diameter, due to the length of the drill point and the thinness of the material. In this case there is no guide at all – the drill will wander and produce a hole to some odd shape. Producing these odd-shaped holes is known as 'lobing'.

The same problem arises with 'snatching' or 'grabbing' – the thinner metal is pushed aside and the drill screws itself through. A further problem associated with this is damage to the metal sheet. A drill pushed with too much force tends to distort the thin sheet initially, rather than cut, and the resulting series of bulges around the holes is unacceptable.

These problems can be overcome by supporting the sheet on a piece of unwanted or waste metal plate. The support prevents distortion and the drill point is guided until the hole is drilled through. There is no problem of breaking through, since the operation is the same as drilling a blind hole, Fig. 7.21. Large-diameter holes can be produced using a trepanning tool as shown in Fig. 7.16.

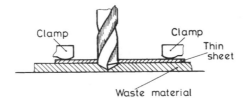

Clamp   Clamp   Thin sheet

Waste material

Figure 7.21 Thin sheet clamped to waste material

7.7 Drilling plastics

Plastics materials cover a wide range of types and applications, which are dealt with in detail in Chapter 14.

In general, plastics materials are easily machined using high-speed-steel cutting tools, although some plastics containing abrasive fillers wear out tools very quickly and the use of diamond tools is essential. Thermosetting plastics can be drilled using standard high-speed-steel twist drills. The chips from thermoplastic materials tend to stick and pack the flutes and cause overheating, which can affect the composition of the material. To prevent this, slow-helix drills with wide highly polished flutes are available. Point thinning can also be carried out to reduce friction and heat at the centre of the drill point. A better finish on breakthrough can be obtained by sharpening the point angle at 90°.

To avoid chipping on breakthrough when drilling the more brittle materials such as Perspex, the material should be held firmly against a solid backing such as a block of hardwood. Use of hardwood prevents damage to the drill point.

Large holes in sheet material can be produced using a trepanning tool.

8 Shaping

A shaping machine is used to produce plain flat surfaces, usually for small quantities or a single workpiece. It is the one basic machine in which the cutting tool reciprocates, or moves backwards and forwards. Since cutting takes place only on the forward stroke, production work on plain flat surfaces is normally carried out on milling machines. However, setting up of the shaping machine and the workpiece is simple, and a number of operations can be carried out at one setting with one single-point cutting tool swung at different angles.

The capacity of a shaping machine is governed by the stroke of the reciprocating ram, the travel of the slide carrying the cutting tool, the area of the table top, and the maximum height between the table top and the cutting tool.

Power to the reciprocating ram may be hydraulic or mechanical.

8.1 The shaping machine

The main elements of a typical mechanical shaping machine are shown in Fig. 8.1.

Column and base

The column and base form the foundation of the complete machine. Both are made from cast iron, designed with thick sections to ensure complete rigidity and freedom from vibration – since the cutting tool reciprocates, shock loads occur at the beginning of each cutting stroke.

The area of the base beneath the table is machined to provide a flat surface for use in supporting the table when taking heavy cuts.

The column is a hollow box section and carries the motor, gearbox, and drive system. At the top of the column is a dovetail slideway in which the ram is guided. An adjustable gib strip is provided to take up wear in the slides.

The front of the column carries guideways for vertical movement of the main slide.

Ram

The ram, mounted on top of the column, is a hollow casting of heavy cross-section to resist shock loading.

A dovetail slide guides the ram in a straight line and prevents it lifting when taking a cut. The pivot bracket, clamped to the underside of the ram by a lockbolt, transmits movements from the main drive. The lockbolt passes through a slot in the ram, which enables the ram position to be altered relative to the work without affecting the length of stroke.

Toolhead

The toolhead, fitted to the front of the ram, is graduated and may be swivelled to any angle up to about 60°. The toolbox slide is carried on a dovetail slideway and is

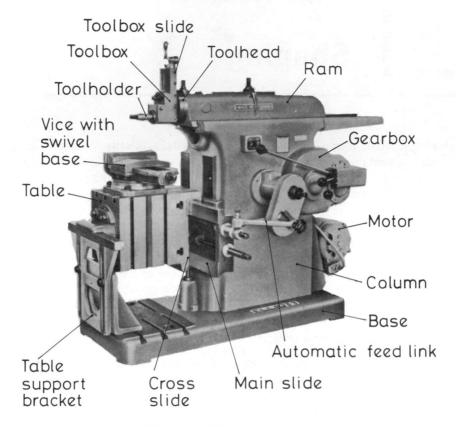

Figure 8.1 Shaping machine

capable of vertical adjustment by a leadscrew through a bronze nut. The leadscrew carries a graduated dial. A toolbox fitted on the front of the toolbox slide can be swivelled to provide clearance for the cutting tool when cutting down a side face. The centre of the toolbox is pivoted to prevent the cutting tool from dragging on the return stroke and carries the toolholder in which the cutting tool is mounted.

Main slide

The main slide, guided on the column guideways, provides vertical movement to the table. Movement is transmitted by a horizontal shaft through bevel gears to a vertical leadscrew. The main slide carries the guideways for the cross slide, which provides movement in a horizontal plane. Movement for the cross slide is through a leadscrew and nut mounted in the main slide.

Any vertical movement of the main slide requires an equal adjustment of the table support, especially when taking heavy cuts. It is usual to adjust the main slide within the range of the toolhead – any cuts can then be taken by adjusting the toolbox slide.

Cross-slide and table

The cross-slide, mounted on the main slide, provides cross traverse to the table.

Attached to the face of the cross slide is a circular centre support, the outer end of which rests on the table support. The table is mounted on the centre support and can be rotated about it up to 45° in each direction. A graduated scale on the front of the table facilitates angular setting. The table is held firmly in the desired position by three clamping bolts at the front. The top and side faces of the table are provided with tee slots to enable vices and workpieces to be clamped to the table surface.

Automatic feed is provided to the cross-slide by means of a pawl through a ratchet fitted at the end of the cross-traverse leadscrew. The pawl is located at the end of a link, the other end of which is adjustable in a slot across the face of a spindle driven through the main drive. Adjustment of the link in the slot allows the pawl to move one, two, three, or four slots in the ratchet, thus providing four rates of feed, Fig. 8.2. The feed operates at the end of each return stroke of the ram. Feed can be disengaged by lifting the pawl clear of the ratchet. The direction of feed can be altered by turning the pawl through 180°.

The cross-feed mechanism is protected from overload or overrun by means of a shear pin fitted in the ratchet arrangement.

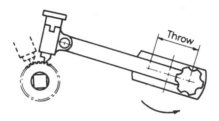

Figure 8.2 Automatic feed to cross slide

Main drive

Power is supplied by an electric motor mounted at the rear of the column. The drive is transmitted to a gearbox on the side of the column by vee belts through a friction clutch operated by a single lever. The gearbox is connected to a large bull gear driven through helical gear teeth on its periphery. The bull gear carries the stroke-adjusting slide across its width. This slide is operated through bevel gears by a shaft which passes through the bull-wheel spindle to the side of the machine. The stroke-adjusting slide carries a pin on which a die is mounted and slides in a slot in the rocking bracket. The top end of the rocking bracket is attached by a pin, through the pivot bracket, which is clamped to the underside of the ram by a lockbolt. The bottom end of the rocking bracket slides on the bottom die, which is attached to a pin secured at the base of the column. The bottom end of the rocking bracket sliding on the bottom die compensates for the change in length of the rocking bracket as it moves across the centre of the bull gear. As the bull gear rotates, the die on the stroke-adjusting slide moves in the slot in the rocking bracket, causing the rocking bracket and the ram attached to its top end to move backwards and forwards. The distance of the die from the centre of the bull wheel determines the length of stroke. Details of the complete drive linkage are shown in Fig. 8.3.

A shaping machine cuts only on the forward stroke of the ram, resulting in an idle return stroke. The rocking bracket is therefore arranged so that the idle return stroke takes a shorter period of time than the forward cutting stroke. This quick-return feature is shown in Fig 8.4. With the stroke-adjusting slide set at radius OA, the die sliding in the rocking bracket moves through ADB on the forward cutting stroke. On the return or idle stroke, the die moves through a distance BCA which is

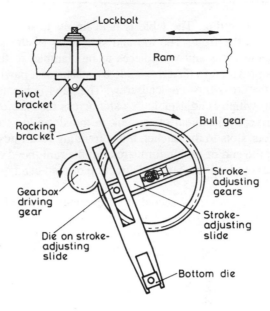

Figure 8.3 Main drive linkage

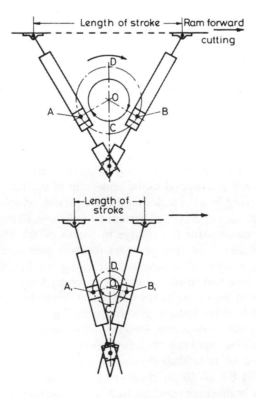

Figure 8.4 Quick-return movement

considerably shorter, and therefore takes less time, than the forward cutting stroke. If the stroke-adjusting slide is set at radius O_1A_1, the length of stroke is shortened, although a similar ratio of forward cutting stroke to idle return stroke still exists.

8.2 Controls

The various controls of a typical shaping machine are shown in Fig. 8.5.

The motor is started and stopped by push buttons (1) located at the side of the column. Drive to the ram is achieved by operating the clutch through a lever (2). The ram can be 'inched' forward or back by slight movement of the clutch lever as an aid to setting the correct length of stroke and position of the ram.

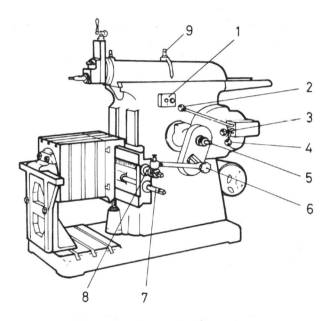

Figure 8.5 Machine controls

Length of stroke is altered by rotation of a shaft (5) after releasing a knurled locknut. This adjustment should be done with the ram stationary. The locknut is then retightened and, by 'inching' the ram, the length of stroke is checked. This can be done in relation to a pointer on the ram. The correct position of the ram can be carried out by 'inching' the ram to its full forward position, slackening the lockbolt (9), and pushing the ram by hand to the correct position. Retighten the lockbolt (9) securely before starting ram movement.

The required speed is selected by operating levers (3 and 4). Lever (3) gives three speeds while lever (4) selects high or low, giving a total range of six speeds, in this case 11, 17, 27, 41, 65, and 101 strokes per minute.

Vertical traverse is carried out through shaft (7), bearing in mind that the table support, where used, will also have to be adjusted accordingly. Cross traverse is carried out through shaft (8).

Feed rate is adjusted by the link (6), and the correct direction is selected by rotating the feed pawl accordingly.

8.3 Shaping operations

Although work can be clamped directly to the table using the tee slots provided, the usual method of workholding is in a machine vice. A vice may have a plane base or a swivel base with a graduated scale to enable workpieces to be swivelled at any required angle, as shown on the typical machine in Fig 8.1. It is essential that the vice is securely clamped to the machine table using the correct size tee bolts – the cutting force on impact at the beginning of the cutting stroke is high, and an insecure vice can be pushed off the end of the table.

Setting the vice

A machine vice must be set up correctly in relation to the machine movements. This is done by fixing a dial indicator to one part of the machine and checking across a parallel held in the vice. If the jaws require to be parallel to the movement of the ram, the dial indicator is attached to the ram and moved, by hand, across the parallel, Fig. 8.6(a). Jaws required to be parallel to the cross movement are checked by attaching the dial indicator to a fixed part of the machine and moving the parallel past the dial indicator, Fig. 8.6(b).

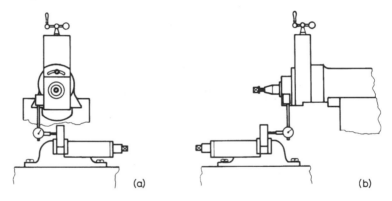

(a)   (b)

Figure 8.6 Setting up the vice

A parallel is used in order to give the same condition as a clamped component. The surface is straight and undamaged, and lengths longer than the vice jaws can be used to give greater accuracy of setting. The vice is adjusted until a constant reading is obtained on the dial indicator, and the vice is then securely clamped down.

Machining sequence

There are two essential requirements when machining any workpiece which has plain surfaces. Firstly the opposite faces must be parallel to each other and square to their adjacent faces, and secondly as many operations as possible should be carried out at a single setting. Bearing these in mind, consider the workpiece shown in Fig. 8.7, which is to be machined on all surfaces.

Set the vice with the jaws in line with the ram movement as previously described. Set the workpiece in the vice with datum face A on two parallels of height such that the workpiece is at least 25 mm above the surface of the vice jaws. Tighten the vice securely, at the same time striking the workpiece with a rawhide mallet to ensure that it is seated on the parallels. This can be checked by moving the ends of the parallels – when no movement is felt, the workpiece is properly seated.

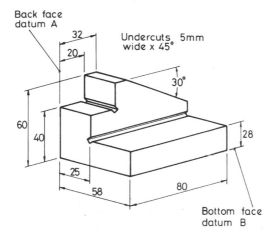

Figure 8.7 Workpiece

Machine face C to clean up, using a round-nose shaper tool as shown in stage 1 of Fig. 8.8.

Release the workpiece and reset it in the vice with face C against the fixed jaw. Tighten the vice, ensuring that the workpiece is seated on the parallels.

Machine face B to clean up, Fig. 8.8 stage 2.

By placing face C against the fixed jaw, which is subject to less wear than the moving jaw, greater accuracy of squareness between faces B and C is achieved.

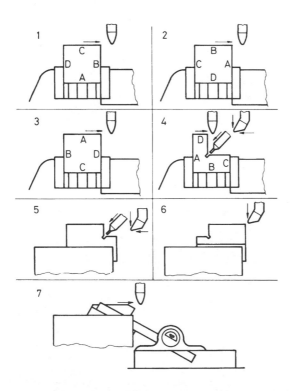

Figure 8.8 Machining sequence

Release the workpiece and reset it in the vice with face B against the fixed jaw and face C seated on the parallels.

Machine face A to achieve the 58 mm dimension, Fig. 8.8 stage 3.

Placing face B against the fixed jaw ensures squareness with face A, and seating the previously machined face C on parallels ensures parallelism of faces A and C. Release the workpiece and reset it with face A against the fixed jaw and face B seated on parallels, i.e. the workpiece is now located with its datum faces against the fixed jaw and on the parallels.

Machine face D to the 60 mm dimension, Fig. 8.8 stage 4.

If the correct height parallels were chosen at the first stage, the step can be machined to the 25 mm and 28 mm dimensions using a cranked tool without the need to reset the workpiece in the vice.

At this same setting, the 45° undercut can be machined by swinging the tool slide at 45° and using a 5 mm wide tool.

Swivel the vice through 180° so that the jaws are in line with the table cross movement and reset the tool slide vertical. Set the workpiece in the vice with datum face A against the fixed jaw and datum face B seated on the parallels. The end of the workpiece is set to protrude beyond the edge of the vice jaws.

Machine the end face to clean up. Machine the step to 20 mm and 40 mm dimensions. Set the tool slide at 45° and, using a 5 mm wide tool, produce the undercut. All these operations are done at the one setting, Fig. 8.8 stage 5.

Release the workpiece and reverse it so that the opposite end is overhanging the end of the vice jaws. Machine to the overall length of 80 mm, Fig. 8.8 stage 6.

The angle is now machined by setting the workpiece in the vice at 30° by means of a protractor set on the worktable, Fig. 8.8 stage 7. If the protractor does not reach the workpiece when set on the table, parallels can be built up and the protractor set on these.

9 Turning

Turning is carried out on a lathe of some description, the type depending on the complexity of the workpiece and the quantity required. All lathes are derived from the centre lathe, so called since the majority of work in the past was done between centres, to ensure concentricity of diameters. This is no longer the case, as accurate methods of workholding are now available.

Centre lathes are made in a variety of sizes and are identified by the maximum size of workpiece which can be machined. The most important capacity is the largest diameter which can be rotated over the bed of the machine, and this is known as the swing. A centre lathe with a swing of 330 mm will accept this diameter of workpiece without it hitting the machine bed. It should be noted that this maximum diameter cannot be accepted over the whole length of the bed, since the cross slide is raised and will therefore reduce the swing. In the case of a 330 mm swing machine, the swing over the cross slide is 210 mm.

The second important capacity is the maximum length of workpiece which can be held between the centres of the machine. A centre lathe with a swing of 330 mm may, for example, accommodate 630 mm between centres.

9.1 Centre-lathe elements

A typical centre lathe showing the main machine elements is shown in Fig. 9.1.

Bed

The lathe bed is the foundation of the complete machine. It is made from cast iron, designed with thick sections to ensure complete rigidity and freedom from vibration. On the top surface, two sets of guideways are provided, each set consisting of an inverted vee and a flat, Fig. 9.2. The arrangement shown may vary on different machines. The outer guideways guide the saddle, and the inner guideways guide the tailstock and keep it in line with the machine spindle. The guideways are hardened and accurately ground.

Two styles of bed are available: a straight bed, where the guideways are continuous over the length of the bed, and a gap bed, where a section of the guideways under the spindle nose can be removed. Removal of this section increases the swing of the lathe, but only for a short distance, Fig. 9.3. For example, the 330 mm swing lathe with a gap bed increases its swing to 480 mm for a length of 115 mm.

The bed is securely bolted to a heavy-gauge steel cabinet containing electrical connections and a tool cupboard, and provides a full-length cutting-fluid and swarf tray.

Headstock

The complete headstock consists of a box-shaped casting rigidly clamped to the guideways of the bed and contains the spindle, gears to provide a range of twelve

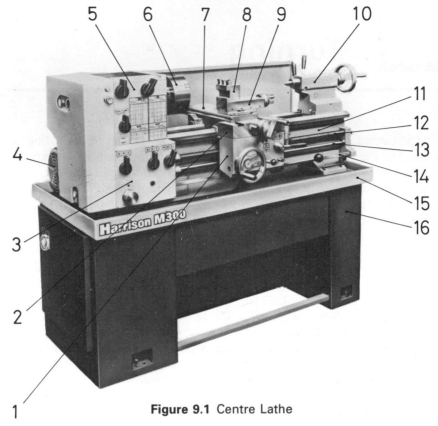

Figure 9.1 Centre Lathe

1. Apron
2. Saddle
3. Gearbox
4. Motor
5. Headstock
6. Chuck
7. Cross slide
8. Toolpost
9. Top slide
10. Tailstock
11. Bed
12. Leadscrew
13. Feed shaft
14. Spindle-control shaft
15. Coolant and swarf tray
16. Cabinet base

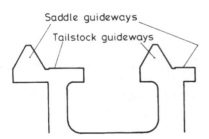
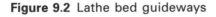

Figure 9.2 Lathe bed guideways

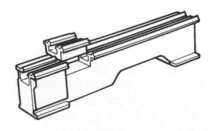

Figure 9.3 Gap bed

spindle speeds, and levers for speed selection. The drive is obtained from the main motor through vee belts and pulleys and a series of gears to the spindle. The speed range is from 40 to 2500 rev/min.

The spindle is supported at each end by precision taper-roller bearings and is bored through to accept bar material. The inside of the spindle nose has a Morse taper to accept centres. The outside of the spindle nose is equipped with means of locating and securing the chuck, faceplate, or other workholding device. The method shown in Fig. 9.4, known as a cam-lock, provides a quick, easy and safe means of securing workholding equipment to the spindle nose. The spindle nose has a taper which locates the workholding device, and on the outside diameter of the spindle nose are three cams which coincide with three holes in the face. The workholding device has three studs containing cut-outs into which the cams lock, Fig. 9.4(a) and (c).

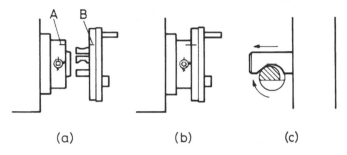

(a)   (b)   (c)

Figure 9.4 Cam-lock spindle nose

To mount a workholding device, ensure that the locating surfaces of both parts are clean. Check that the index line on each cam lines up with the corresponding line on the spindle nose, Fig. 9.4(a). Mount the workholding device on the spindle nose, ensuring that the scribed reference lines A and B on the spindle nose and the workholding device line up. These lines assist subsequent remounting. Lock each cam by turning clockwise, using the key provided. For correct locking conditions, each cam must tighten with its index line between the two vee marks on the spindle nose, Fig. 9.4(b); if this does not happen, do not continue but inform your supervisor or instructor who can then carry out the necessary adjustment. Since each workholding device is adjusted to suit a particular spindle, it is not advisable to interchange spindle-mounted equipment between lathes.

Removal of equipment is carried out by rotating each cam anticlockwise until the index lines coincide and then pulling the equipment away from the spindle nose.

The gearbox, fitted on the lower side of the headstock, provides the range of feeds to the saddle and cross slide through the feed shaft, and the screw-cutting range through the leadscrew. By selecting the appropriate combination of lever positions in accordance with a table on the machine, a wide range of feed rates and thread pitches can be obtained.

Tailstock

The function of the tailstock is to hold a centre when turning between centres, or to act as a support at the end of long workpieces. Alternatively, the tailstock is used to hold drills and reamers when producing holes.

The tailstock can be moved on its guideways along the length of the bed and locked in any position. The quill contains a Morse-taper bore to accommodate centres, chucks, drills, and reamers and is graduated on its outer top surface for use when drilling to depth. It can be fed in or out by means of the handwheel at the rear. Positive locking of the quill is carried out by means of a handle operating an eccentric pin.

Saddle

The saddle rests on top of the bed and is guided by two guideways which, for stability, are the two furthest apart. Accurate movement is thus maintained relative to the centre line of the spindle and tailstock for the complete length of the bed. The top surface contains the dovetail slideway into which the cross-slide is located and the cross-slide leadscrew, complete with handwheel and graduated dial, Fig. 9.5.

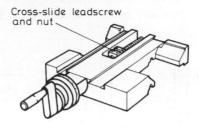

Figure 9.5 Saddle

Cross-slide

Mounted in the dovetail slideway on the top surface of the saddle, the cross-slide moves at right angles to the centre line of the machine spindle. Adjustment for wear is provided by a tapered gib strip, which can be pushed further into the slide and slideway by the screw as wear takes place. Attached to the underside of the cross-slide is the leadscrew nut through which movement is transmitted from the leadscrew. Power feed is available to the cross-slide.

The top surface contains a radial tee slot into which two tee bolts are fitted. The central spigot locates the slideway for the top slide, which can be rotated and clamped at any angle by means of the tee bolts. Graduations are provided for this purpose, Fig. 9.6.

On the lathe shown, external dovetails are provided along each side of the cross-slide, for quick accurate attachment of rear-mounting accessories.

Top slide

The top slide shown in Fig. 9.6, often referred to as the compound slide, fits on its slideway and can be adjusted for wear by means of a gib strip and adjusting screws. Movement is transmitted by the leadscrew through a nut on the slideway. A toolpost, usually four-way hand-indexing, is located on the top surface and can be locked in the desired position by the locking handle. Movement of this slide is usually quite short, 92 mm on the machine illustrated, and only hand feed is available. Used in conjunction with the swivel base, it is used to turn short tapers.

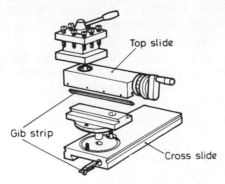

Figure 9.6 Cross slide and top slide

Apron

The apron is attached to the underside of the saddle at the front of the machine and contains the gears for transmission of movement from the leadscrew and feed shaft. Sixteen feed rates from 0.03 to 1 mm per revolution are provided.

On the front are the handles to engage and disengage the leadscrew and feed shaft. Also mounted on the front is the handwheel for longitudinal traverse of the carriage along the bed, this movement being transmitted through gears to a rack fixed on the underside of the bed.

The complete assembly of apron, saddle, and slides is known as the carriage. The spindle control on the apron is operated by lifting for spindle reverse, lowering for spindle forward, and mid position for stop.

9.2 Centre-lathe controls

The various controls of a typical centre lathe are shown in Fig. 9.7.

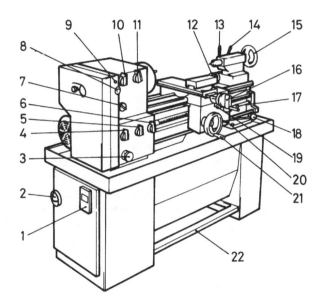

Figure 9.7 Machine controls

1. Coolant-pump starter	9. 'Supply on' lamp	17. Threadcutting
2. Mains isolator	10. Speed selector	engagement
3. Feed-selector dial	11. Speed selector	18. Spindle-control lever
4. Feed-selector handle	12. Top-slide traverse handle	19. Feed-axis selector
5. Feed-selector handle	13. Quill lock	20. Feed-engage lever
6. Feed-selector handle	14. Tailstock clamp	21. Longitudinal-traverse
7. Feed-direction selector	15. Quill-traverse handwheel	handwheel
8. Emergency stop	16. Cross-traverse handle	22. Emergency-stop and
		brake pedal

Before starting the machine, ensure that the feed-engage lever (20) and the thread-cutting lever (17) are in the disengaged position.

Select the feed axis required, i.e. longitudinal travel of carriage or cross-slide, by means of the apron push–pull knob (19).

Select the direction of feed by means of selector handle (7).

Select the feed rate required by referring to the charts on the headstock and selecting the appropriate position of selector dial (3) and handles (4), (5), and (6).

Select the spindle speed by means of selector handles (10) and (11).

Switch on the main electrical supply at the mains isolator (2).

Start the spindle by lifting the spindle-control lever (18) for reverse or lowering it for forward. The mid position is 'stop'.

Start and stop the feed motion as required by means of the feed-engage lever (20).

Do not attempt to change speeds and feeds when the spindle is running – always stop the machine first.

Stopping the machine

The machine can be stopped by returning the spindle-control lever (18) to its central stop position. Alternatively, press the emergency-stop push button (8) or depress the full-length foot brake pedal (22).

9.3 Workholding

Workpieces can be held in a centre lathe by a variety of methods depending on the shape and the operation being carried out.

The most common method of holding work is in a chuck mounted on the end of the spindle. Several types of chuck are available, the most common being the three-jaw self-centring chuck, the four-jaw independent chuck, and the collet chuck.

Three-jaw self-centring scroll chuck

This chuck, Fig. 9.8, is used to hold circular or hexagonal workpieces and is available in sizes from 100 mm to 600 mm. It operates by means of a pinion engaging in a gear on the front of which is a scroll, all encased in the chuck body. The chuck jaws, which are numbered and must be inserted in the correct order, have

Figure 9.8 Three-jaw chuck

teeth which engage in the scroll and are guided in a slot in the face of the chuck body. As the pinion is rotated by a chuck key, the scroll rotates, causing all three jaws to move simultaneously and automatically centre the work.

Two sets of jaws are usually supplied: those which grip externally while turning, facing, and boring, Fig. 9.9(a), and those which grip internally while the outside diameter or face is machined, Fig. 9.9(b).

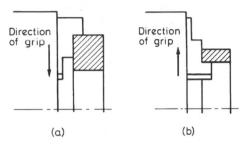

Figure 9.9 (a) Outside, and (b) inside jaws

Four-jaw independent chuck

The four-jaw independent chuck, Fig. 9.10, is used to hold square, rectangular, and irregular shaped work which cannot be held in the three-jaw self-centring type. It is available in sizes from 150 mm to 1060 mm. As the name implies, each jaw is operated independently by means of a screw – the jaws do not move simultaneously.

Figure 9.10 Four-jaw independent chuck

Although the jaws are numbered and must be replaced in the appropriate slot, they are reversible, due to the single-screw operation.

Concentric rings are machined in the front face to aid setting up the work, and tee slots are sometimes provided on the front face for additional clamping or packing of awkward workpieces.

Collet chuck

This type of chuck, Fig. 9.11, fits on the spindle nose and is convenient for bar and the smaller-diameter workpieces. Having fewer moving parts than the moving-jaw types makes it more accurate. It is more compact and does not have the same overhang from the spindle nose, and work can be machined up to the front of the collet. All-round gripping of the component makes it ideal for holding tube and thin-walled workpieces which tend to collapse in the three- or four-jaw chucks.

Figure 9.11 Collet chuck

In the model shown in Fig. 9.12, each collet is produced with a number of blades and will accommodate slight size variation up to 3 mm.

Figure 9.12 Multi-size collet

Chuck keys

Accidents occur when chuck keys are left in the chuck and the machine is inadvertently switched on.

No matter for how short a period, *never* leave the chuck key in the chuck. Safety chuck keys, Fig. 9.13, are now available which are spring-loaded and, if left in position, pop out and fall from the chuck.

Faceplate

The faceplate, Fig. 9.14, is used for workpieces which cannot be easily held by any of the other methods. When fixed to the machine, the face is square to the machine-spindle centre line. A number of slots are provided in the face for clamping purposes. Workpieces can be clamped to the faceplate surface but, where there is a risk of machining the faceplate, the workpiece must be raised from the surface on parallels before clamping. Positioning of the workpiece depends upon its shape and the accuracy required.

Figure 9.13 Safety chuck key

Flat plates which require a number of holes are easily positioned by marking out the hole positions and using a centre drill in a drilling machine to centre each position. A centre in the tailstock is then used to locate the centre position and hold the workpiece against the faceplate while clamping is carried out, Fig. 9.14.

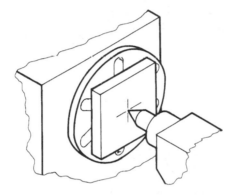

Figure 9.14 Locating workpiece on faceplate

Workpieces which already contain a hole which is to be enlarged, e.g. cored holes in a casting, can be marked out to produce a box in the correct position, the sides of which are the same length as the diameter of the required hole. Roughly positioned and lightly clamped, the workpiece can be set accurately using a scriber in a surface gauge resting on the cross-slide surface. The faceplate is rotated by hand and the workpiece is tapped until all of the scribed lines are the same height, indicating that the hole is on centre, Fig. 9.15. The workpiece is then securely clamped.

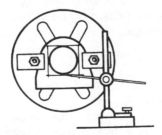

Figure 9.15 Setting workpiece on a faceplate

Accurate positioning of holes in a plate can be done with the aid of toolmaker's buttons. These consist of a hardened and ground steel bush of known diameter with the ends ground square and a flanged screw. The required hole positions are marked out and a hole is drilled and tapped to suit the screw. The accuracy of the drilled and tapped hole is not important, as there is plenty of clearance between the screw and the bore for the button to be moved about. The button is then held on the work by the screw and is accurately positioned by measuring across the outside of adjacent buttons using a micrometer. The buttons are adjusted until the required distance is reached and are then securely tightened by means of the screw, Fig. 9.16(a). The centre distance of the hole to be produced is $x - d$.

The workpiece is then lightly clamped to the faceplate and is accurately positioned using a dial indicator on the button, Fig. 9.16(b). The workpiece is securely clamped and the button is removed. The hole is drilled and bored in the knowledge that it is in the correct position. This is repeated for the remaining holes.

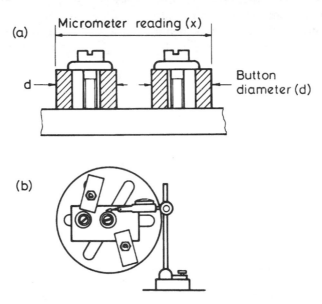

Figure 9.16 Using toolmaker's buttons

Precautions for faceplate work: When the workpiece has been clamped, check each nut and screw to ensure it is tight. Turn the faceplate by hand and check that all bolts and clamps are clear of the bed, cross slide, or toolpost. To ensure this, avoid using excessively long clamping bolts. Check for 'out of balance' of the faceplate – a counterbalance may be required.

Centres

Components having a number of diameters which are required to be concentric can be machined between centres. A centre is inserted in the spindle nose, using a reducing bush supplied for this purpose. This centre rotates with the spindle and workpiece and is referred to as a 'live' centre. A centre inserted in the tailstock is fixed, does not rotate, and is referred to as a 'dead' centre. Great care must be taken to prevent overheating of 'dead' centres due to lack of lubrication or too high a pressure. Keep the centre well lubricated with grease, and do not overtighten the tailstock.

In order to drive the workpiece, a work-driver plate must be mounted on the spindle nose and the drive is completed by attaching a work carrier to the workpiece, Fig. 9.17.

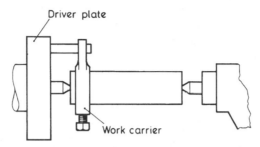

Figure 9.17 Workpiece between centres

Where the size of workpiece requires greater pressure and where considerable time is required for the operation, 'dead' centres will 'burn out', i.e. overheat and the point wear out. To overcome this, live or rotating tailstock centres are available, the centres of which run in bearings which will withstand high pressures without overheating.

Tailstock centres are often required for long work which is held in the chuck but requires support owing to its length.

Steadies

If unsupported, long slender work may tend to be pushed aside by the forces of cutting. To overcome this, a two-point travelling steady is used which provides support to the workpiece opposite the tool as cutting is carried out along the length of the work, Fig. 9.18.

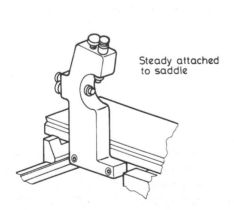

Figure 9.18 Two-point travelling steady

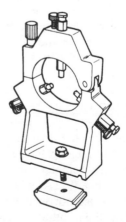

Figure 9.19 Three-point fixed steady

Work of a larger diameter than can be accepted through the machine spindle and yet requiring work to be carried out at one end can be supported using a three-point fixed steady. This steady is clamped to the machine bed and the points are adjusted so that the workpiece is running true to the spindle centre line before the machining operation is carried out, Fig. 9.19.

Mandrel

Work which has a finished bore and requires the outside to be turned concentric to it can be mounted on a mandrel. The mandrel is then put between centres and the work is machined as already described for 'between-centres' work.

A mandrel, Fig. 9.20, is a hardened and ground bar with centres in each end and a flat machined at one end to accept the work carrier. The diameter is tapered over its length, usually about 0.25 mm for every 150 mm length. When the work is pushed on, this slight taper is enough to hold and drive the work during the machining operation.

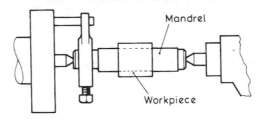

Figure 9.20 Mandrel

The flat for the carrier is machined on the end having the larger diameter, so that the carrier does not have to be removed to load and unload workpieces.

9.4 Centre-lathe operations

Turning

Accurate turning of plain diameters and faces can be simply carried out on a centre lathe. Wherever possible, diameters should be turned using the carriage movement, as the straightness of the bed guideways ensures parallelism of the workpiece and power feed can be used. Avoid using the top slide for parallel diameters, since it is adjustable for angle and difficult to replace exactly on zero without the use of a dial indicator. It has also to be hand fed.

When a number of diameters are to be turned on a workpiece, they should be produced at one setting without removing the workpiece from the chuck, in order to maintain concentricity between them. Accuracy is lost each time the workpiece is removed and put back in the chuck. Accurate sizes can be produced by measuring the workpiece when the final size is almost reached, then using the graduated dial on the handwheel to remove the required amount.

Where only diameters are being turned and a square shoulder is required, a knife tool is used, Fig. 9.21A, which cuts in the direction shown. Where facing and turning are being carried out in the same operation, a turning and facing tool is used, Fig. 9.21B. The slight radius on the nose produces a better surface finish, but the radius will be reproduced at the shoulder.

Where a relief or undercut at the shoulder is required, e.g. where a thread cannot be cut right up to the shoulder, an undercut tool is used. This tool is ground to the correct width, the face parallel to the work axis, and is fed in the direction shown, Fig. 9.21C.

Work produced from bar can be cut to length in the lathe, an operation known as 'parting off'. The face of the parting-off tool is ground at a slight angle, so that the workpiece is severed cleanly from the bar, Fig. 9.21D.

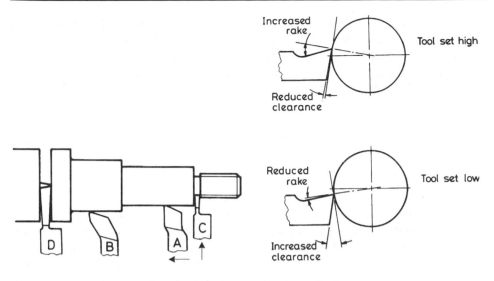

Figure 9.21 Turning-tool applications   **Figure 9.22** Effect of tool set above and below centre

It is essential that all cutting tools used on a lathe be set on the centre of the workpiece. A tool set too high reduces the clearance and will rub, while one set too low reduces the rake angle, Fig. 9.22. Cutting tools can be set relative to a centre inserted in the tailstock and then be raised or lowered using suitable thicknesses of packing, Fig. 9.23. A good stock of varying thickness of packing should be available which, when finished with, should always be returned for future use.

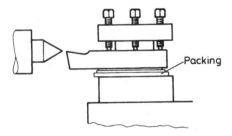

Figure 9.23 Setting tool on centre

Drilling

Drilling is carried out on a lathe by holding the drill in a chuck or mounting it directly in the quill of the tailstock, which contains a Morse taper for this purpose. As with all drilling operations, some guide is required to enable the drill to start central, and a centre drill, Fig. 9.24, is commonly used.

Figure 9.24 Centre drill

Centre drills are available in various sizes, and their purpose is to produce centres in a workpiece for turning between centres. Due to its rigid design, a centre drill is convenient in providing a suitable guide to start the drill in the centre of the bar.

When using a centre drill, great care must be taken to prevent breakage of the small point which, because of its size, does not have deep flutes to accommodate swarf. Feed in gently a short distance at a time, using the tailstock handwheel, winding the drill out frequently to remove swarf before it packs the flute and snaps off the point. Use high spindle speeds for the small point diameter. The centre drill should be fed in just deep enough to give the drill a start.

Drilling is then carried out to the required depth, which can be measured by means of the graduations on the quill. Relieve the drill frequently, to prevent swarf packing the flutes.

Reaming

Holes requiring a more accurate size and better surface finish than can be achieved with a drill can be finished by reaming. The hole is drilled about 0.4 mm smaller than required, followed by the reamer using a spindle speed approximately half that used for drilling. A reamer will follow the hole already drilled, and consequently any error in concentricity or alignment of the hole axis will not be corrected by the reamer. Where accurate concentricity and alignment are required, the hole should be drilled a few millimetres undersize, bored to within 0.4 mm of the required size, correcting any error, and finally reamed to achieve the finished size.

Boring

As already stated, boring can be used to correct errors in concentricity and alignment of a previously drilled hole. The hole can be finished to size by boring without the use of a reamer, as would be the case when producing non-standard diameters for which a reamer was not available. Boring is also used to produce a recess which may not be practical by drilling and reaming, Fig. 9.25.

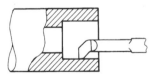

Figure 9.25 Boring tool

A boring tool must be smaller than the bore it is producing, and this invariably results in a thin flexible tool. For this reason it is not usually possible to take deep cuts, and care must be taken to avoid vibration. In selecting a boring tool, choose the thickest one which will enter the hole, to ensure maximum rigidity. Ensure also that adequate secondary clearance is provided in relation to the size of bore being produced, as shown in Fig. 6.4.

9.5 Taper turning

The method used to turn a taper depends upon the angle of taper, its length, and the number of workpieces to be machined. Three methods are commonly used: with a form tool, with the top or compound slide, and with a taper-turning attachment.

Form tool

Short tapers of any angle can be produced by grinding the required angle on the cutting tool, Fig. 9.26. The cutting tool is then fed into the work until the desired length of taper is produced.

This method is normally used for short tapers such as chamfers, both internal and external. The long cutting edge required by the long tapers has a tendency to chatter, producing a bad surface finish.

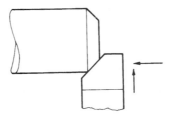

Figure 9.26 Angle form tool

Top or compound slide

Taper turning can be carried out from the top slide by swivelling it to half the included angle required on the work, Fig. 9.27. Graduations are provided on the base plate, but any accurate angle must be determined by trial and error. To do this, set the top slide by means of the graduations, take a trial cut, and measure the angle. Adjust if necessary, take a second cut, and remeasure. When the correct angle is obtained, ensure that the clamping nuts are securely tightened.

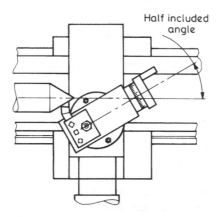

Figure 9.27 Top slide set at half included angle

Turning the angle is done by winding the top-slide handle by hand. The tool will feed at the angle to which the top slide is set. After the first cut, the tool is returned to its starting position by rewinding the top slide. The feed for the second cut is achieved by moving the cross-slide.

This method can be used for any angle, internal or external, but the length is restricted by the amount of travel available on the top slide.

Taper-turning attachment

Taper-turning attachments can be fitted at the rear of the cross-slide and can be used to turn included angles up to 20° over a length of around 250 mm, both internally and externally.

A plan view of a typical taper-turning attachment is shown in Fig. 9.28. The guide bar, which swivels about its centre, is mounted on a base plate which carries the graduations. The base plate is attached to the connecting rod, which passes through a hole in the clamp bracket where it is held tightly by a clamping screw. The clamp bracket is clamped to the bed of the machine. Thus the guide bar, base plate, connecting rod, and clamp bracket are securely fixed to each other and to the machine bed.

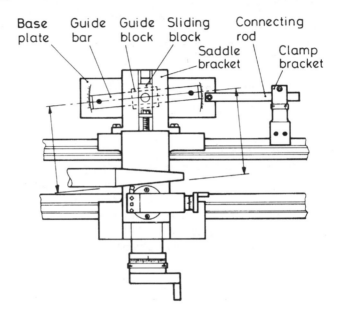

Figure 9.28 Taper-turning attachment

The guide block slides on the guide bar and is located in the sliding block by a spigot. This gives a solid location and at the same time allows the guide block to take up the angle of the guide bar.

The sliding block is attached to the end of the cross-slide leadscrew and is guided in a bracket which is bolted to the rear face of the saddle.

It can therefore be seen that, if the carriage is traversed along the bed and the guide bar remains stationary (i.e. clamped to the bed), the sliding block can only push or pull the cross-slide leadscrew. For this movement to be transmitted to the cross slide and so to the cutting tool, a special leadscrew is required, Fig. 9.29. The front end of the leadscrew has a spline which slides up the inside of the handwheel spindle. When the sliding block pushes the leadscrew, the leadscrew moves back and, since it passes through the leadscrew nut which in turn is screwed to the cross-slide, the cross-slide and the cutting tool mounted on it will also move back, pushing the spline up the inside of the handwheel spindle.

By this method, a cut can be put on merely by rotating the handwheel, driving through the spline to the leadscrew and nut without interfering with the taper-turning attachment. To revert to a normal operating condition, the connecting rod is

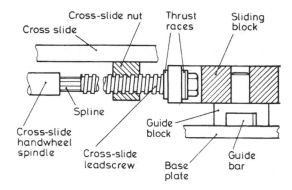

Figure 9.29 Cross-slide leadscrew for taper-turning attachment

unclamped and the clamp bracket removed and, since the complete attachment moves with the carriage, the cross slide can then be used in the normal way.

9.6 Screw-cutting

The thread now standardised in British industry is the ISO metric thread, ISO being the International Organisation for Standardisation. Terminology of this thread is shown in Appendix 1.

The ISO metric thread has a 60° truncated form, i.e. the thread does not come to a sharp point but has a flat crest. The root of the thread also has a small flat.

A single-point tool sharpened as shown in Fig. 9.30 produces the thread angle and the flat at the root, the major diameter being produced at the turning stage. To cut an accurate thread requires a definite relationship between the rotation of the work in the spindle and the longitudinal movement of the carriage by means of the leadscrew. All modern centre lathes have a gearbox through which a wide range of pitches can be obtained by referring to a chart on the machine and turning a few knobs.

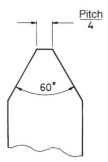
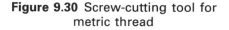
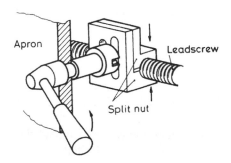

Figure 9.30 Screw-cutting tool for metric thread

Figure 9.31 Split nut for screw cutting

The longitudinal travel of the carriage is obtained from the leadscrew through a split nut housed in the apron and operated by a lever on the apron front, Fig. 9.31. By closing the split nut, the drive can be started at any position.

The position of engagement of the split nut on the leadscrew for each cut is important in order that the tool will travel along the same path as the previous cut. To achieve this accuracy of engagement, a thread indicator dial is fitted at the end of the apron, Fig. 9.32. The dial is mounted on a spindle at the opposite end of which is a gear in mesh with the leadscrew. These gears are interchangeable, are stored on the spindle, and are selected by referring to a chart on the unit. They are arranged to give a multiple of the pitch required, relative to the 6 mm pitch of the leadscrew.

The chart shows the gear used for a particular pitch of thread and the numbers on the thread indicator dial at which the split nut may be engaged. To cope with the different diameters of gears, the unit pivots and is locked in position when the gear is in mesh. To avoid unnecessary wear, the unit is pivoted back out of mesh when not in use for screw-cutting.

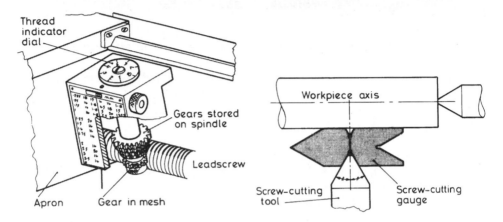

Figure 9.32 Thread indicator dial   **Figure 9.33** Screw-cutting gauge and positioning of tool

Method

Having turned the workpiece to the correct diameter, the following procedure should be followed. This procedure is for screw-cutting a right-hand external metric thread on a machine having a metric leadscrew.

1. Carefully grind the tool to 60° with the aid of a screw-cutting gauge, Fig. 9.33, leaving a flat on the tool nose.
2. Mount the tool in the toolpost on the centre of the workpiece.
3. Set the tool relative to the axis of the workpiece, using the screw-cutting gauge, Fig. 9.33.
4. Calculate the required thread depth.
5. Select the required pitch.
6. Select the correct gear on the thread indicator dial and mesh with the leadscrew.
7. Engage a slow spindle speed.
8. Start the machine.
9. Wind in the cross-slide until the tool just touches the outside of the workpiece and move the carriage so that the tool is clear of the end of the workpiece.
10. Stop the machine.
11. Set the dial on the cross-slide to zero.
12. Restart the machine.
13. Wind the cross-slide to give a small cut of 0.05 mm.

14. Wait until the appropriate number on the thread indicator dial comes round to the mark and engage the split nut.
15. Take a trial cut.
16. When the end of the workpiece is reached, unwind the cross-slide to remove the tool from the work and disengage the split nut. This is done in one movement. At all times during screw-cutting, one hand should be resting on the cross-slide handwheel, the other on the split-nut lever.
17. Stop the machine.
18. Check the thread to make sure the correct pitch has been cut.
19. Rewind the carriage to the starting point.
20. Restart the machine.
21. Rewind the cross-slide back to the original graduation and put on a further cut.
22. Wait for the correct number on the thread indicator dial, engage the split nut, and repeat until the final depth is reached.

Depending upon the accuracy of thread required, final checking should be carried out by means of a gauge or by checking against a nut or the mating workpiece.

Internal threads are cut in exactly the same manner, except that the tool is similar to a boring tool ground to give a 60° thread form.

Left-hand threads are produced in the same manner by reversing the rotation of the leadscrew and starting from the opposite end of the workpiece.

Imperial threads

Imperial threads are designated not by their pitch but by the number of threads per inch (t.p.i.). The leadscrew of a metric centre lathe has a pitch of 6 mm and, since the number of threads per inch cannot be arranged as a multiple of the leadscrew pitch, the split nut, once it is engaged, must never be disengaged during the thread-cutting operation. This also means that the thread indicator dial is of no use when cutting imperial threads on a metric lathe.

The procedure when cutting imperial threads on a metric lathe is the same as before up to the point when the split nut is disengaged and a trial cut taken.

16. When the end of the workpiece is reached, withdraw the tool and stop the machine but do not disengage the split nut.
17. Reverse the spindle direction so that the carriage moves back to the starting point.
18. Stop the machine, put on a further cut, and restart the machine spindle in a forward direction.
19. Repeat until the thread has been cut to size before disengaging the split nut.

10 Surface grinding

Surface grinding is used to produce flat accurate surfaces and can be carried out on all materials, hard or soft. There may be no other way of removing metal from a hardened workpiece. It is normally considered a finishing operation, but large machines are used in place of milling and shaping machines to remove large amounts of material.

A typical surface grinder is shown in Fig. 10.1 and uses a 300 mm diameter by 25 mm wide grinding wheel. The reciprocating table and cross-slide movements are hydraulically operated, although alternative hand operation is provided.

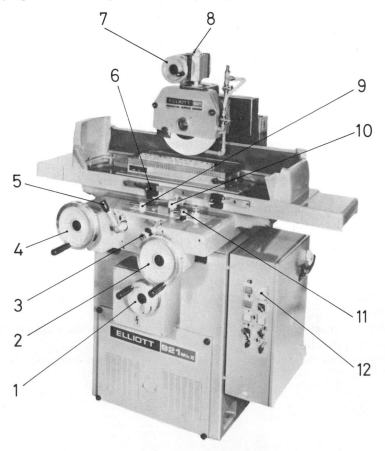

Figure 10.1 Surface grinding machine

The capacity of such a machine is the maximum length and width of surface which can be ground, in this case 500 mm × 200 mm, and the maximum height which can go under a grinding wheel of maximum diameter. Using a 300 mm diameter wheel, the maximum height of workpiece on the machine shown is 400 mm.

10.1 Elements of a surface-grinding machine

The main elements of a typical surface-grinding machine are shown in Fig. 10.2.

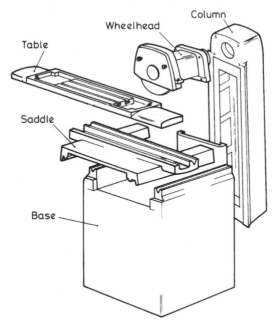

Figure 10.2 Main elements of surface grinder

Base

The base is a heavily ribbed box-section casting to ensure rigidity and complete freedom from vibration. The bottom of the base houses the hydraulic pump and fluid reservoir. At the rear of the base is a vertical dovetail slideway which guides the column. Two vee slideways on top of the base guide the saddle and are widely spaced to maintain accuracy and rigidity.

Column

The column, guided on a dovetail slide, carries the wheelhead at its top end and contains the motor and belt drive to the wheel spindle. The column and wheelhead are raised and lowered through a screw and nut from a handwheel on the front of the machine. A telescopic guard is fitted to prevent grinding dust coming between the slide surfaces.

Wheelhead

The wheelhead carries the wheel spindle, which is mounted in precision bearings. The complete grinding-wheel collet assembly is fitted on a taper on the end of the spindle. Drive to the spindle is by vee belt and pulley from the motor mounted in the bottom of the column.

Saddle

The saddle is fitted on top of the base in the two vee slideways and provides the cross-traverse movement. The cross traverse can be applied automatically in continuous or incremental feed by hydraulic power or, alternatively, with a manually operated handwheel. The automatic cross movement is infinitely variable up to a maximum of 10 mm. The increment of cross movement is applied at each end of the table stroke, resulting in complete grinding of the workpiece surface in the manner shown schematically in Fig. 10.3. The top surface carries a vee-and-flat slideway to guide the table at right angles to the saddle movement.

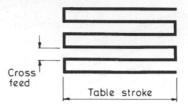

Figure 10.3 Schematic diagram of cross-feed movement to table stroke

Table

The table is guided by the vee-and-flat slideway on the saddle and can be manually operated with a handwheel. Automatic reciprocation of the table is transmitted through a hydraulic cylinder at infinitely variable speeds from 0.6 to 30m/min. Reversal of the table movement is achieved automatically by trip dogs operating a direction-reversing valve. The trip dogs can be set to give the required length of table stroke and position of reversal.

A simplified diagram of the hydraulic circuit is shown in Fig. 10.4. When the direction-reversing valve A is in the position shown, the sliding valve B moves to the right. This allows hydraulic fluid into the left of the table cylinder which is attached to the saddle. The fluid moves the piston, and the table attached to it, to the right.

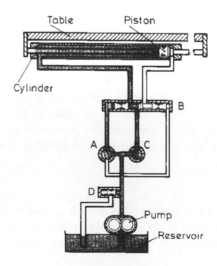

Figure 10.4 Simplified diagram of table hydraulics

When the end of the table stroke is reached, the trip dog moves the direction-reversing valve, causing fluid to move the sliding valve to the left. This allows fluid into the right of the table cylinder, moving the piston and table to the left. Thus automatic continuous reciprocating movement of the table is achieved. Other connections into this circuit are made to give automatic cross movement at the end of each table stroke.

Control valve C meters the amount of fluid reaching the cylinder and so controls the speed of the table movement. Relief valve D allows any pressure build-up to be released. This prevents mechanical damage in the event of accidental overload or jamming of the table, as the fluid is merely returned to the reservoir.

A series of tee slots is provided on the top table surface to enable clamping of workpieces or workholding equipment.

10.2 Controls

Controls of a typical surface grinder are shown in Fig. 10.1.

Handwheel (1) raises and lowers the column. By lowering the column, a cut is put on by the wheel. Since the accuracy of the workpiece depends on how much metal is removed, the graduations on this handwheel represent very small increments of movement, in this case 0.0025 mm.

Handwheel (2) provides cross movement of the saddle, graduations on this handwheel representing increments of 0.01 mm.

Handwheel (4) is used to reciprocate the table by hand.

Length of stroke and position of table reversal are controlled by trip dogs (6) striking the direction-reversing-valve lever (10).

The table-speed control knob (11) can be adjusted to give infinitely variable speeds from 0.6 to 30 m/min.

Lever (9) is used to select continuous cross feed or incremental feed at the end of each table stroke.

Where continuous cross feed is selected, lever (5) controls the speed, which is infinitely variable from 0 to 5 m/min.

The rate of incremental feed is controlled by lever (3) and is infinitely variable from 0.28 to 10 mm.

The switch panel at the right side of the machine controls the motors for the hydraulic pump, wheel spindle, cutting fluid, etc. and carries the mains isolator and a large mushroom-headed stop button (12).

10.3 Workholding

The basic method of workholding in surface grinding is the permanent-magnet chuck, used to hold workpieces having flat surfaces. These chucks will not hold non-magnetic materials such as the non-ferrous range. The complete chuck consists of a top plate containing inserts separated from the top plate by a non-magnetic epoxy-resin filler, a non-magnetic case, a moving grid containing the permanent magnets insulated from the grid and magnetised vertically, and a base plate, Fig. 10.5.

The principle upon which permanent-magnet devices operate is to establish the magnetic lines of force or flux from the permanent magnets through the workpiece when switched on, and to divert or 'short circuit' the flux when switched off. This is achieved by moving the magnets in line with the top plate and so completing the circuit to the insert, grid, and base plate through the workpiece Fig. 10.6(a). For a workpiece to be gripped, it is therefore necessary for it to bridge the top plate and an

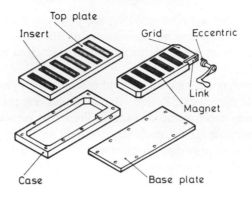

Figure 10.5 Permanent-magnet chuck

insert. To switch off, the magnets are moved out of line with the top plate, diverting the flux so that the circuit is completed not through the workpiece but through the top plate, insert, and base plate, Fig. 10.6(b). The workpiece is thus deprived of flux and is released.

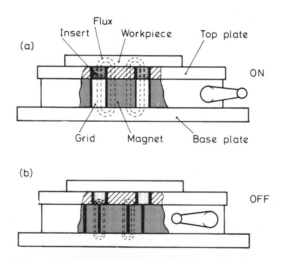

Figure 10.6 Permanent-magnet chuck showing flux lines in (a) ON and (b) OFF positions

Other methods of workholding are used when the shape or the material from which the workpiece is made does not allow direct holding on the permanent-magnet chuck. However, the devices used to hold the workpieces are invariably themselves held on the permanent-magnet chuck.

Vices are used to hold workpieces, but it should be remembered that grinding is usually a finishing operation and so any vice used should be accurate and if possible kept only for use in grinding. Care must also be taken to avoid distortion of the component, as this will be reflected in the finished workpiece.

Surfaces required to be ground at right angles can be clamped to the upright surface of an angle plate.

Vee blocks are used to hold circular workpieces.

10.4 Grinding wheels

All machining operations are potentially dangerous – lack of understanding or undue care have resulted in many accidents. The use of grinding wheels, also known as abrasive wheels, which are described fully at the end of this chapter, is potentially one of the most dangerous for two reasons.

- A grinding wheel is made of small abrasive particles held together by a bonding material. Compared with metal it is extremely fragile.
- Grinding wheels are run at high speeds. A 300 mm diameter wheel is run at about 2000 rev/min, giving a speed at the diameter of almost 1900 m/min. Compare this with a piece of steel of the same diameter being cut with a high-speed-steel cutting tool on a lathe at 30m/min.

To minimise the risks in the use of grinding wheels, special training must be given under the Abrasive Wheels Regulations 1970. The Abrasive Wheels Regulations 1970 state that

'No person shall mount an abrasive wheel unless he

a) has been trained in accordance with the schedule to these Regulations,
b) is competent to carry out that duty,
c) has been appointed and an entry made and dated in the official register maintained by his Company.'

Although most of the Abrasive Wheels Regulations 1970 has been replaced by the Provision and Use of Work Equipment Regulations 1992 (PUWER) the detailed training requirements have not, and will continue to apply.

No attempt is to be made here to cover the requirements for instruction under the Regulations, merely to draw attention to them.

Dressing

A grinding wheel is made up of a large number of tiny teeth. The teeth are formed by the tiny grains of hard abrasive, held together by a bonding material. As with any other metal-cutting operation, the 'teeth' or grains must be kept sharp. To some extent a grinding wheel is self-sharpening. The ideal situation during grinding is that, as the grains which are cutting become blunt, greater force is exerted which tears the blunt grains from the bonding material, exposing fresh sharp ones.

It follows that, when grinding a hard material, the grains become blunt quickly and will require to be torn from the bonding material quickly. To allow this to take place, less bonding material is used to hold the grains, the grains tear away easily when blunt, and the wheel is referred to as 'soft'. The opposite is true when grinding a soft material: the grains do not blunt so readily and can be held in position longer, and therefore more bonding material is used. These wheels are referred to as 'hard'.

Sharpness and trueness of the face can be achieved by 'dressing' with an industrial diamond. The diamond, held in a suitable holder, is positioned on the chuck under the grinding wheel. The wheel is then lowered until it touches the diamond, whereupon the diamond is moved across the surface of the wheel using the cross-traverse movement, Fig. 10.7. The wheel is then lowered a little and the operation is repeated until all the worn grains have been torn out and fresh ones exposed and the face is flat and true.

The machine shown in Fig. 10.1 has a built-in dressing attachment mounted above the wheel. By turning the graduated dial (8) in Fig. 10.1, the diamond is lowered in

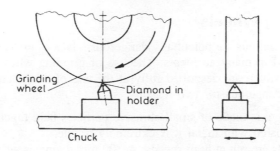

Figure 10.7 Dressing the grinding wheel using a diamond

contact with the top of the wheel and is traversed across the wheel surface using the handwheel (7). An attachment of this kind saves time through not having to position and remove the diamond for each dressing nor having to lower the wheel for dressing and to raise it again to continue grinding.

Balancing

It is impossible to produce good-quality work on any grinding machine if the wheel is out of balance, thus setting up vibrations through the spindle.

The wheel is mounted on a collet, the complete assembly being removable from the spindle, Fig. 10.8. The wheel spigot upon which the wheel is located has a taper bore to locate on the spindle nose. The wheel flange locates on the wheel spigot and is held by three screws which, when tightened, securely hold the wheel between the two surfaces. On the outer face of the wheel flange is an annular dovetail groove which holds balance masses. These masses can be locked in any position round the groove.

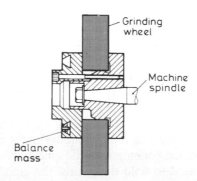

Figure 10.8 Wheel-collet assembly

When the wheel is mounted correctly and the periphery has been dressed, the collet assembly is removed from the spindle. The balance masses are then removed. A balancing arbor is inserted in the bore. The balancing arbor has a taper identical to that on the spindle and has equal-size parallel diameters at each end, Fig. 10.9(a). This assembly is placed on a balancing stand, Fig. 10.9(b), which has previously been set level. The wheel is allowed to roll on the knife edges and is left until it comes to rest, which it will do with the heaviest portion at the bottom.

A chalk mark is made at the top of the wheel, opposite the heaviest portion. The masses are then replaced in a position opposite to each other and at right angles to

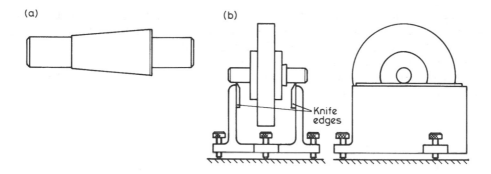

Figure 10.9 (a) Balancing arbor and (b) stand

the chalk mark. The masses can now be moved equally a little way towards the light portion and locked. The wheel is then rolled and allowed to come to rest. If the same heavy portion again comes at the bottom, both masses are moved a little closer to the light portion and again the wheel is allowed to roll and come to rest. This process is repeated until the wheel will stop in any position and show no tendency to roll along the balancing stand. The wheel is then ready for use and should be replaced on the spindle in the correct manner.

Guarding

Due to the fragile nature of a grinding wheel and the high speeds at which it runs, it is possible for a wheel to burst. Under the regulations, all grinding wheels must be adequately guarded and the guard be fitted at all times before the wheel is run.

On surface grinders, the guard consists of a plate which encloses the front of the wheel and retains fragments, so protecting the operator should the wheel burst.

Bonded-abrasive grinding wheels

A bonded-abrasive wheel consists of two main essentials: the abrasive, which does the actual cutting, and the bonding material or bond, which holds the abrasive together and forms the wheel shape, Fig. 10.10.

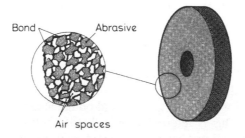

Figure 10.10 Abrasive-wheel features

The ideal cutting condition of an abrasive wheel is that, when they have done their work and become dull, the abrasive grains fracture or are released from the bonding material to expose new sharp cutting grains in their place. This is repeated continuously during cutting and gives the wheel a self-sharpening effect.

If the grains are released before they have done their work and become dull, the wheel wears rapidly and is said to act 'soft'.

When the strength of the bonding material is too great to allow the grains to be released when they have become dull, the wheel is said to act 'hard' and is recognisable by a glazed and shiny appearance on the cutting face.

In order to grind a range of materials efficiently under a variety of cutting conditions, different abrasives of varying grain size are arranged within different bonding materials to give a number of features or characteristics.

Abrasives

Two abrasive materials are in general use: aluminium oxide and silicon carbide.

Aluminium oxide is produced by fusing bauxite in special electric furnaces. When cool, it is crushed to produce shaped particles which are then graded in a series of grain sizes. Owing to its tough nature, aluminium oxide is suitable for grinding metals of high tensile strength, such as steel. This abrasive is designated by the letter A and is available under the trade names 'Aloxite', 'Alundum', and 'Bauxilite'.

Silicon carbide is made by the chemical reaction, in an electric furnace, of high-carbon coke and pure silica sand, with small amounts of salt and sawdust added to assist the reactions. The resultant mass is crushed to give grains of the correct shape which are then graded in a series of grain sizes. This is a hard and brittle abrasive and is most efficient for grinding materials of low tensile strength, such as cast iron, brass, copper, aluminium, and glass. This abrasive is designated by the letter C and is available under the trade names 'Crystolon' and 'Unirundum'.

Grain size

The size of the abrasive particles is indicated by a number representing the number of openings per linear inch in the screen used for sizing. The standard grain sizes, from the coarsest to the finest, are 8, 10, 12, 14, 16, 20, 24, 30, 36, 46, 54, 60, 70, 80, 90, 100, 120, 150, 180, 220, 240, 280, 320, 400, 500 , and 600. The sizes most widely used range from 10 to 120.

The grain size used affects the amount of material which will be removed and the final surface roughness. For rough grinding requiring a large amount of material removal, a large grain size, e.g. 10, is used, resulting in a rough surface. If a fine surface is required, a small grain size, e.g. 120, is used. The small grains will not, of course, be able to remove the same large amount of material as the large grains, but they produce a fine surface finish.

Grade

This is the strength of the bond holding the abrasive grains in place. It is a measure of the amount of bond present.

More bond material will have a greater hold on the abrasive grains, which will be less readily released. This is referred to as a 'hard' grade.

Less bond material will not have such a great hold on the abrasive grains, which will then be released more readily. This is referred to as a 'soft' grade.

When grinding a hard material, the abrasive grains dull more quickly than when grinding a soft material. This means that the grains must be released more quickly, to expose new sharp grains, so a wheel having a soft grade is used, which allows easy release of the grains as they become dull.

When grinding soft materials, the grains are not required to work as hard, do not dull quickly, and therefore do not need to be released as readily. A hard-grade wheel can be used which will retain the grains for a longer period.

Hence the common saying in industry: 'A hard wheel for soft materials; a soft wheel for hard materials.'

Grades are designated from the softest to the hardest by letters A to Z.

Structure

The structure of an abrasive wheel is determined by the proportions and arrangement of the abrasive and bond. The grains are spaced to leave smaller or larger air spaces or pores between them.

Wheels where the spacing is wide, and large air spaces exist, are classified as having an 'open' structure. The large air spaces provide clearance for metal chips as they are removed from the work. Since larger chips are removed from the softer materials, open-structure wheels are used on such materials.

In grinding conditions where a large area of contact exists between wheel and work, a greater amount of heat is generated. This is often the case in surface and internal grinding. As the open-structure wheel has large air spaces, fewer grains are in contact and therefore less heat is generated. This type of wheel used in these conditions is said to have a free cool-cutting action.

Wheels where the spacing is closer and smaller air spaces exist are classified as having a 'dense' structure. When hard metals are being ground, small chips are produced which do not need a large clearance in the wheel. Dense-structure wheels are therefore used for these types of material.

Due to the close spacing, dense-structure wheels generate more heat, which can result in burning the surface of the metal being ground. This is recognisable by brown patches on the metal surface. If soft materials are ground with a dense-structure wheel, the clearance for metal chips is insufficient, the air spaces clog with metal, and the wheel is said to be 'loaded'.

Structure is designated by a series of numbers from 0 to 14, indicating the densest to the most open spacing.

Bond

Various bonding materials are used, depending chiefly on the amount of material which is to be removed. This is related to the abrasive-wheel speed.

The most widely used bonds are vitrified, resinoid, and rubber.

Vitrified: The majority of wheels have vitrified bonds. The abrasive grains are mixed in the correct proportions with clay and fusible materials. This mixture is then pressed in moulds to produce the correct wheel shape. The wheels are passed through drying rooms, to remove any moisture, before being fired in a kiln. During the firing process, the clay and fusible materials melt to form bonds between adjacent abrasive grains. On cooling and solidification, a 'glass-like' material is formed.

This type of bond is designated by the letter V. Vitrified wheels are porous, strong, and unaffected by water, oils, and ordinary temperature conditions. In general, surface speeds do not exceed 1950 m/min.

Resinoid: To produce resinoid-bond wheels, the abrasive grains are mixed with a thermosetting synthetic resin, moulded to shape, and cured. This bond is very hard and strong, and wheels can be run at surface speeds from 2850 to 4800 m/min. These high surface speeds give rapid metal removal, due to the greater number of abrasive particles cutting in any given time, and wheels with this bond are therefore suitable where high rates of metal removal are required and are ideal for work in foundries and steel mills.

Resinoid-bonded wheels can be made extremely thin, can be used with safety at high speeds, and are ideal as cut-off wheels for cutting metal bars, tubes, etc. This type of bond is designated by the letter B.

Where a higher-strength bond is required, with a degree of flexibility, an open-weave fabric reinforcement is incorporated in the resin, designated by the letters BF.

Rubber: To produce rubber-bond wheels, the abrasive grains are mixed with rubber and vulcanising agents running between heated rolls. After rolling to thickness, the wheels are cut to the correct diameter and then vulcanised or 'cured' – undergoing a chemical reaction in which the molecules of rubber are interlinked using heat and pressure, usually with the aid of sulphur, to give a high degree of resilience.

Very thin wheels can be made by this process, because of the elasticity of the material, and are ideal for cut-off wheels, where they are run at high surface speeds between 3000 and 4800 m/min. Rubber-bond wheels are also used for control wheels in centreless-grinding machines.

This type of bond is designated by the letter R.

Characteristics

The characteristics already described are designated by up to seven symbols (three of which are optional) arranged in the following order:

0 – type of abrasive (optional), manufacturer's own symbol
1 – nature of abrasive
2 – grain size
3 – grade
4 – structure (optional)
5 – nature of bond
6 – type of bond (optional), manufacturer's own symbol

Symbols 1 to 5 are selected from the standard symbols set out in BS 4481: part 1 and shown in Fig. 10.11. These standard symbols are normally marked on the wheel by the manufacturer or, in the case of small wheels, on the box or package in which they are contained.

10.5 Surface-grinding operations

Surface grinding is used to produce flat accurate surfaces. This can be illustrated by considering the grinding of all surfaces of the component shown in Fig. 10.12.

It is important to first establish the datum faces A and B from which all faces are then ground.

Clamp face B against an angle plate supported on a parallel and grind face A to clean up, Fig. 10.13 stage 1. Reclamp with face A against the angle plate, again supported on a parallel, and grind face B to clean up, Fig. 10.13 stage 2. This ensures that datums A and B are square to each other.

It is usual practice to fit a 'fence' to the rear of a magnetic chuck and to grind its face using the side of the grinding wheel. This face is then parallel to the table movement. The faces of a workpiece pushed against the fence can also be ground parallel using the side of the grinding wheel.

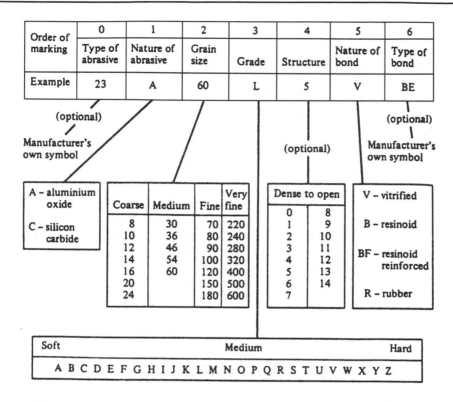

Figure 10.11 Standard symbols for marking of a grinding wheel

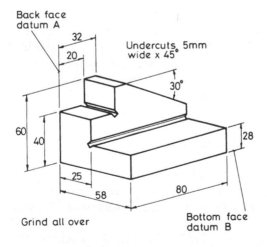

Figure 10.12 Workpiece

Place the workpiece on the magnetic chuck on face B, with face A against the 'fence'.

Grind the step face to 28 mm thickness, taking care to avoid hitting the adjacent face with the side of the grinding wheel.

Raise the wheelhead and grind the top face to the 60 mm dimension, Fig. 10.13 stage 3.

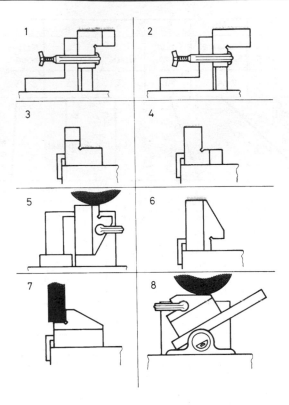

Figure 10.13 Sequence of operations

Reset the workpiece on the magnetic chuck on face A, with face B against the 'fence'. Grind the top face to the 58 mm dimension. Lower the wheelhead and grind the step face to the 25 mm dimension, taking care to avoid hitting the adjacent face with the side of the grinding wheel, Fig. 10.13 stage 4.

Clamp face A against an angle plate and set face B vertical, using a square.

Grind the end face to clean up, Fig. 10.13 stage 5.

Reset the workpiece on the magnetic chuck on the end face just ground.

Grind the opposite end face to the 80 mm dimension, Fig. 10.13 stage 6.

Reset the workpiece on the magnetic chuck on face B with the end face against the 'fence'.

Grind the step face to the 40 mm dimension and the adjacent face to the 20 mm dimension, using the side of the grinding wheel, Fig. 10.13 stage 7.

Clamp face A against the angle plate and tilted at 30°, using a protractor under face B.

Grind the 30° angle face, Fig. 10.13 stage 8, to achieve the 32 mm dimension.

11 Milling

Milling is the machining of a surface using a cutter which has a number of teeth. The surface produced may be plain or, by using additional equipment or special cutters, formed surfaces may be produced.

There are many types and sizes of milling machines, but the most versatile in common use in the majority of workshops is the knee-and-column type, so called because the spindle is fixed in the column or main body and the table arrangement, mounted on a knee, is capable of movement in the longitudinal, transverse, and vertical directions.

Knee-and-column machines are subdivided into the following models:

- plain horizontal, with the spindle located horizontally;
- universal, which is similar to the plain horizontal but equipped with a swivelling table for use when cutting helical grooves;
- vertical, with the spindle located vertically.

Typical plain horizontal and vertical knee-and-column milling machines are shown in Figs 11.1 and 11.2.

Figure 11.1 Horizontal milling machine

Figure 11.2 Vertical milling machine

The capacity of these machines is identified by the size of the working surfaces of the table, the length of travel of the longitudinal, transverse, and vertical movements, and the maximum distance from spindle to table surface on the horizontal model or from spindle to column on the vertical model.

11.1 Milling-machine elements

The main elements of a typical knee-and-column horizontal milling machine are shown in Fig. 11.3. The elements of a vertical machine are the same except that the spindle head is mounted at the top of the column, as shown in Fig. 11.4.

Column and base

The column and base form the foundation of the complete machine. Both are made from cast iron, designed with thick sections to ensure complete rigidity and freedom from vibration. The base, upon which the column is mounted, is also the cutting-fluid reservoir and contains the pump to circulate the fluid to the cutting area.

The column contains the spindle, accurately located in precision bearings. The spindle is driven through a gearbox from a vee-belt drive from the electric motor housed at the base of the column. The gearbox enables a range of spindle speeds to be selected. In the model shown, twelve spindle speeds from 32 to 1400rev/min are available. The front of the column carries the guideways upon which the knee is located and guided in a vertical direction.

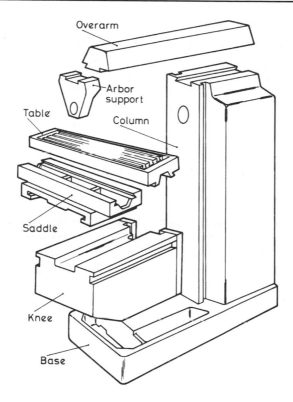

Figure 11.3 Main machine elements of horizontal milling machine

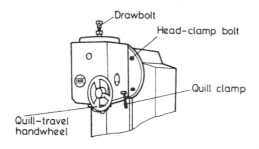

Figure 11.4 Top of column of vertical milling machine

Knee

The knee, mounted on the column guideways, provides the vertical movement of the table.

Power feed is available, through a gearbox mounted on the side, from a separate built-in motor, providing a range of twelve feed rates from 6 to 250 mm/min. Drive is through a leadscrew, whose bottom end is fixed to the machine base. Provision is made to raise and lower the knee by hand through a leadscrew and nut operated by a handwheel at the front. The knee has guideways on its top surface giving full-width support to the saddle and guiding it in a transverse direction.

A lock is provided to clamp the knee in any vertical position on the column.

Saddle

The saddle, mounted on the knee guideways, provides the transverse movement of the table.

Power feed is provided through the gearbox on the knee. A range of twelve feeds is available, from 12 to 500 mm/min. Alternative hand movement is provided through a leadscrew and nut by a handwheel at the front of the knee.

Clamping of the saddle to the knee is achieved by two clamps on the side of the saddle.

The saddle has dovetail guideways on its upper surface, at right angles to the knee guideways, to provide a guide to the table in a longitudinal direction.

Table

The table provides the surface upon which all workpieces and workholding equipment are located and clamped. A series of tee slots is provided for this purpose. The dovetail guides on the undersurface locate in the guideways on the saddle, giving straight-line movement to the table in a longitudinal direction at right angles to the saddle movement.

Power feed is provided from the knee gearbox, through the saddle, to the table leadscrew. Alternative hand feed is provided by a handwheel at each end of the table. Stops at the front of the table can be set to disengage the longitudinal feed automatically in each direction.

Spindle

The spindle, accurately mounted in precision bearings, provides the drive for the milling cutters. Cutters can be mounted straight on the spindle nose or in cutter-holding devices which in turn are mounted in the spindle, held in position by a drawbolt passing through the hollow spindle. Spindles of milling machines have a standard spindle nose, shown in Fig. 11.5, to allow for easy interchange of cutters and cutter-holding devices. The bore of the nose is tapered to provide accurate location, the angle of taper being 16° 36'. The diameter of the taper depends on the size of the machine and may be 30, 40, or 50 IST (International Standard Taper). Due to their steepness of angle, these tapers – known as non-stick or self-releasing – cannot be relied upon to transmit the drive to the cutter or cutter-holding device. Two driving keys are provided to transmit the drive.

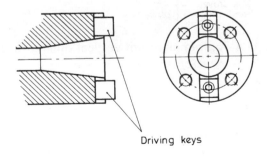

Driving keys

Figure 11.5 Standard milling-machine spindle nose

Cutters which are mounted directly on the spindle nose are located on a centring arbor, and four tapped holes are provided to hold the cutter in position. The two keys again provide the means of transmitting the drive.

The spindle of a horizontal machine is fixed and cannot be adjusted in an axial direction, i.e. along its axis. On vertical machines, provision is made for axial movement, which is controlled by a handwheel on the spindle head. The spindle runs in a quill which is moved through a rack and pinion in the same way as a drilling-machine spindle (see Fig. 7.2). A locking bolt is provided to lock the quill in any position along its operating length.

Overarm and arbor support

The majority of cutters used on horizontal machines are held on an arbor which is located and held in the spindle. Due to the length of the arbors used, support is required at the outer end to prevent deflection when cutting takes place. Support is provided by an arbor-support bracket, clamped to an overarm which is mounted on top of the column in a dovetail slide. The overarm is adjustable in or out for different lengths of arbor, or can be fully pushed in when arbor support is not required. Two clamping bolts are provided to lock the overarm in any position. The arbor support is located in the overarm dovetail and is locked by means of its clamping bolt. A solid bearing is provided in which the arbor runs during spindle rotation.

11.2 Controls

The various controls of a typical horizontal milling machine are shown in Fig. 11.6.These are identical to those of a vertical machine.

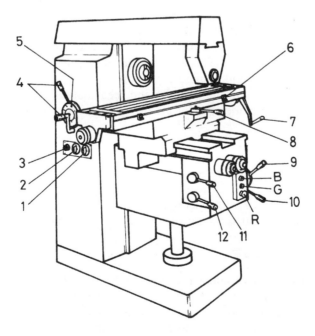

Figure 11.6 Milling-machine controls

Spindle speeds are selected through the levers (4), and the speed is indicated on the change dial (5). The speeds must not be changed while the machine is running. An 'inching' button (3) is situated below the gear-change panel and, if depressed, 'inches' the spindle and enables the gears to slide into place when a speed change is being carried out. Alongside the 'inching' button is the switch for controlling the cutting-fluid pump (1) and one for controlling the direction of spindle rotation (2). The feed rates are selected by the lever (9) and are indicated on the feed-rate dial.

To engage the longitudinal table feed, lever (8) is moved in the required direction – right for right feed, left for left feed. Adjustable trip dogs (6) are provided to disengage the feed movement at any point within the traverse range. Limit stops are incorporated to disengage all feed movements in the extreme position, to prevent damage to the machine in the event of a trip dog being missed.

To engage cross or vertical traverse, lever (12) is moved up or down. The feed can then be engaged by moving lever (11) in the required direction. With cross traverse selected, movement of lever (11) upwards produces in-feed of the saddle, moving it downwards produces out-feed of the saddle. With vertical traverse selected, movement of lever (11) upwards produces up-feed to the knee, moving it downwards produces down-feed to the knee.

Rapid traverse in any of the above feed directions is engaged by an upward pull of lever (10). Rapid traverse continues as long as upward pressure is applied. When released, the lever will drop into the disengaged position. Alternative hand feed is provided by means of a single crank handle (7), which is engaged by slight pressure towards the machine. Spring ejectors disengage the handle on completion of the operation, for safety purposes – i.e. the handle will not fly round when feed or rapid traverse is engaged. The single crank handle is interchangeable on table, saddle, and knee movements.

Starting and stopping the machine

The switch panel, situated on the front of the knee, contains a black button (B) to start the feed motor, Fig. 11.6. This is provided to facilitate setting up when feed movements are required without spindle rotation.

The green button (G) starts the spindle and feed motors, while the mushroom-headed red button (R) provides the means of stopping the machine.

11.3 Milling cutters

There are many different types of milling cutter available, and for convenience they can be classified according to the method of mounting: those with a central hole for mounting on an arbor, those with a screwed shank for holding in a special chuck, and the large facing cutters which mount directly on to the spindle nose.

Arbor-mounted types

Cylindrical cutter: This cutter has teeth on the periphery only, and is used to produce flat surfaces parallel to the axis of the cutter, Fig. 11.7(a). The teeth are helical, enabling each tooth to take a cut gradually, reducing shock and minimising chatter. Cylindrical cutters are made in a variety of diameters and lengths up to 160 mm diameter × 160 mm long.

Side-and-face cutter: This cutter has teeth on the periphery or face and on both sides. It is used to produce steps, cutting on the face and side simultaneously, Fig. 11.7(b), or for producing slots. The use of these cutters in pairs with their sides cutting is known as straddle milling, Fig. 11.7(c). The teeth are straight on cutters up

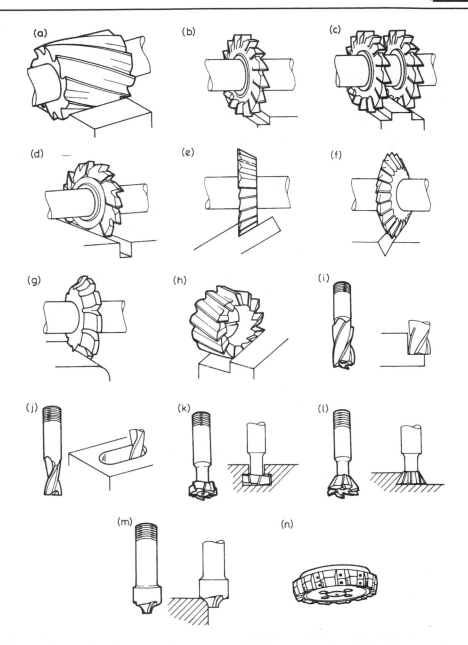

Figure 11.7 Milling cutters: (a) cylindrical cutter, (b) side-and-face cutter, (c) straddle milling, (d) staggered-tooth side-and-face cutter, (e) and (f) angle-milling cutters, (g) single corner-rounding cutter, (h) shell end-milling cutter, (i) end-milling cutter, (j) slot drill, (k) tee-slot cutter, (l) dovetail cutter, (m) corner-rounding cutter, (n) face-milling cutter

to 20 mm wide but are helical above this thickness. Side-and-face cutters are available in a variety of sizes up to 200 mm diameter and 32 mm wide.

A staggered-tooth side-and-face cutter, also having teeth on the periphery and on both sides, is designed for deep-slotting operations. In order to reduce chatter and provide maximum chip clearance, the teeth are alternately right-hand and left-hand helix, and each alternate side tooth is removed, Fig. 11.7(d). Staggered-tooth cutters are available in the same sizes as the plain side-and-face type.

Angle-milling cutter: Angle cutters are available, made with single angle or double angle with teeth on the angled surfaces. A single-angle cutter also has teeth on the flat side. They are used on angle faces or for producing a chamfer on the edge of the workpiece, Fig. 11.7(e) and (f).

Single-angle cutters are available with angles of 60° to 85° in 5° steps, and the double-angle cutters with 45°, 60°, and 90° included angle.

Single corner-rounding cutter: This cutter has a concave quarter circle on one side and is used to produce a corner radius on the edge of the workpiece, Fig. 11.7(g). Cutters are available with a variety of corner radii from 1.5 mm to 20 mm.

Shell end-milling cutter: More often referred to as a 'shell end mill', Fig. 11.7(h), this cutter has teeth cut on the circumference and on one end. The tooth end is recessed to receive a screwhead for holding the cutter on an arbor. A key slot on the back face provides the drive from two keys in the arbor. The teeth are helical and the cutter is used for work of a larger size than can be efficiently handled by the ordinary end mill. A range of sizes is available from 40 mm to 160 mm diameter.

Screwed-shank types

End-milling cutter: This cutter has helical teeth on the circumference and teeth on one end and is used for light operations such as milling slots, profiling, and facing narrow surfaces, Fig. 11.7(i). The end teeth are not cut to the centre, so this cutter cannot be fed in a direction along its own axis. It is available in a variety of sizes up to 50 mm diameter.

Slot drill: Usually having two helical teeth cut in the circumference and two teeth on the end, cut to the centre, this cutter can be fed along its own axis in the same way as a drill. It is used to produce keyways and blind slots with the cutter sunk into the material like a drill and fed longitudinally the length of the keyway or slot, cutting on its circumference, Fig. 11.7(j). It is available in a variety of sizes up to 50 mm diameter.

Tee-slot cutter: Designed for milling tee slots in machine tables, this cutter has teeth on its circumference and on both sides. To reduce chatter and provide maximum chip clearance, the teeth are alternately right-hand and left-hand helix, and each alternate side tooth is removed. To produce a tee slot, the groove is first cut using a side-and-face cutter, end mill, or slot drill and finally the wide slot at the bottom is cut using a tee-slot cutter, Fig. 11.7(k). The shank is reduced to clear the initial groove. This cutter is available for standard tee slots to suit bolt sizes up to 24 mm.

Dovetail cutter: Designed for milling dovetail slides of machines, this cutter has teeth on its angle face and on the end face. To produce a dovetail slide, a step is machined to the correct depth and width; the angle is then finally machined using the dovetail cutter, Fig. 11.7(l). It is available in a variety of sizes up to 38 mm diameter, with 45° and 60° angles.

Corner-rounding cutter: Designed to produce a radius along the edge of the workpiece, this cutter has a quarter circle cut in the outer edge, Fig. 11.7(m). It is available in a variety of sizes with corresponding radii up to a maximum radius of 12 mm.

Direct-mounted types

Face-milling cutter: More usually referred to as a 'face mill', this cutter is used to face large surfaces. The cutter consists of a tough steel body with high-speed-steel or tungsten-carbide cutting edges in the form of inserts clamped in their correct position, Fig. 11.7(n). This construction results in a cheaper cutter than would be the case if the complete cutter were made from an expensive cutting-tool material – it also facilitates the replacement of one insert in the event of damage to a single cutting edge. It is available in a variety of sizes from 100 mm to 450 mm diameter, the larger sizes being used only on the biggest machines, since the power requirement for such a cutter may be as high as 75kW.

11.4 Cutter mounting

Arbor-mounted cutters

Standard arbor: Milling cutters having a hole through the centre are mounted on an arbor. The standard arbor used in horizontal milling machines is shown in Fig. 11.8. One end has an international taper to suit the machine spindle, for location. A threaded hole in the end provides the means of holding the arbor in position, by means of a drawbolt through the machine spindle. The flange contains two key slots to provide the drive from two keys on the spindle nose. The long diameter is a standard size, to suit the hole size of the cutter, and the thread carries the arbor nut to clamp the cutter. A keyway is cut along the length of this diameter into which a key is fitted, to provide a drive and prevent the cutter slipping when taking heavy cuts. To position the cutter along the length of the arbor, spacing collars are used. These are available in a variety of lengths, with the ends ground flat and parallel. Towards the end, a larger bush is positioned. This has an outside diameter to suit the bearing of the arbor support and is known as the 'running bush'.

To mount the arbor, the taper is inserted in the machine spindle, ensuring that the surfaces are free of all dirt and metal cuttings. The flange key slots are located in the

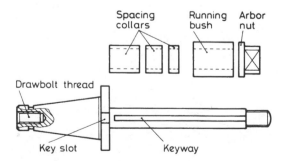

Figure 11.8 Standard milling-machine arbor

spindle keys, and the arbor is securely held by the drawbolt. Spacing collars are slipped on the arbor, again ensuring that all faces are clean and free from dirt and metal cuttings. The cutter is positioned and spacing collars are added, together with the running bush, to make up the length of the arbor. The arbor nut is then screwed in position – hand-tight only.

The arbor support is now positioned on the overarm so that it is central on the running bush and is then clamped in position. The arbor nut can now be tightened with the appropriate spanner. Never tighten the arbor nut without the arbor support in position, as the arbor can be bent.

To prevent deflection of the arbor during heavy cutting operations, it is sometimes necessary to mount a second arbor support nearer the spindle nose. The cutter is then positioned between the two supports.

Stub arbor: Cutters which are used close to the spindle, such as shell end mills, are mounted on a stub arbor, Fig. 11.9. This arbor is located, held, and driven in the spindle in the same way as a standard arbor. The cutter is located on a spigot or stub and is held in position by a large flanged screw. Two keys on the arbor provide the drive through key slots in the back face of the cutter.

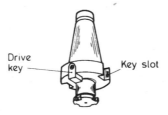

Figure 11.9 Milling-machine stub arbor

Screwed-shank cutters

Cutters having screwed shanks are mounted in a special chuck, shown in Fig. 11.10. The collet, which is split along the length of its front end and has a short taper at the front, is internally threaded at its rear end. Collets of different sizes are available to suit the shank diameter of the cutter used. The collet is inserted into the locking

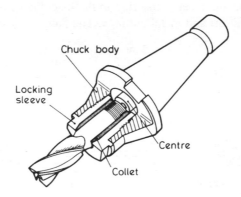

Figure 11.10 Milling chuck for screwed-shank cutters

sleeve and the assembly is screwed into the chuck body until the flange almost meets the end face of the body.

The cutter is inserted and screwed into the collet until it locates on the centre inside the chuck body and becomes tight. The centre anchors the end of the cutter and ensures rigidity and true running. A spanner is used to give the locking sleeve a final tighten.

The cutter cannot push in or pull out during the cutting operation. Any tendency of the cutter to turn during cutting tightens the collet still further and increases its grip on the cutter shank. This type of collet chuck is located, held, and driven in the machine spindle in the same way as the previously mentioned types.

Direct-mounted cutters

Large face mills are mounted directly on the spindle nose. To ensure correct location and concentricity, a centring arbor with the appropriate international taper is held in the spindle by the drawbar. The diameter on the end of the centring arbor locates the cutter, which is driven by the spindle keys through a key slot in the back face of the cutter. The cutter is held in position by four screws direct into the spindle nose, Fig. 11.11.

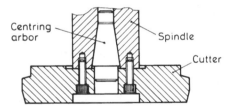

Figure 11.11 Direct-mounted cutter

11.5 Workholding

The simplest method of holding a workpiece for milling is to clamp it directly to the worktable. Adequate tee slots are provided for this purpose. Care should be taken to avoid machining the table – if necessary, the workpiece should be raised on a pair of parallels. Clamping should be carried out in the manner already described in the chapter on drilling (Section 7.3).

Vice

A vice is the most versatile piece of equipment for holding workpieces. It must be positioned in the manner described in Chapter 8, to ensure accurate alignment with machine movements.

Rotary table

A rotary table, Fig. 11.12, is used where part of the surface being machined is of a circular nature. In this case, the table is moved to bring the workpiece in the correct position under the cutter. The workpiece is then rotated past the cutter to produce a circular profile.

The rotary table consists of a base with lugs to clamp to the machine table, inside which a circular table is rotated by means of a handwheel at the front. A dial round the periphery of the table is graduated, usually in degrees. Some models have a

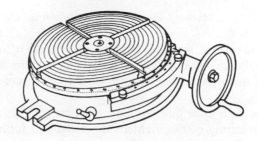

Figure 11.12 Rotary table

vernier scale fitted to give more accurate readings, in some cases as small as 1 minute of arc.

The circular table is provided with tee slots, to enable clamping of the workpiece. A central hole enables setting of the workpiece about the centre of rotation and enables the rotary table to be set central with the machine spindle. Concentric circles on the table surface are provided to aid the initial setting of the workpiece roughly central.

11.6 Milling operations

Consider machining the workpiece shown in Fig. 11.13, using in the first instance a horizontal milling machine and secondly producing the same workpiece using a vertical milling machine. The two essential requirements stated in the chapter on shaping must also be met in milling operations – namely, opposite faces parallel to each other and square with their adjacent faces, and as many operations as possible done at a single setting. The most convenient method of holding such a workpiece is to grip it in a machine vice.

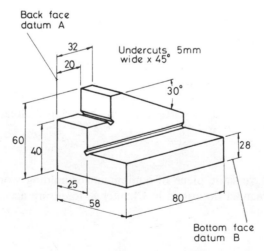

Figure 11.13 Workpiece to be machined

The machine vice is set relative to the machine movements using a parallel gripped in the jaws and is checked by means of a dial indicator attached to a fixed part of the machine, as previously described for the shaping machine (Section 8.3). The machine vice, having been satisfactorily set up, is securely clamped to the machine table.

Operations on a horizontal machine

Set the block in the vice on parallels and ensure that at least 32 mm is protruding above the vice jaws. Tighten the vice, ensuring that the workpiece is seated on the parallels.

Machine face C to clean up, using a cylindrical cutter, Fig. 11.14 stage 1.

Release the workpiece and reset with face C against the fixed jaw. Tighten the vice, ensuring that the workpiece is seated on the parallels.

Machine face B to clean up, Fig. 11.14 stage 2. This ensures squareness of faces B and C.

Release the workpiece and reset with face B against the fixed jaw and face C seated on the parallels.

Machine face A to achieve the 58 mm dimension, Fig. 11.14 stage 3. This ensures parallelism of faces A and C and squareness with face B.

Release the workpiece and reset with face A against the fixed jaw and face B seated on the parallels, i.e. the workpiece is located with its datum faces against the fixed jaw and on the parallels.

Machine face D to achieve the 60 mm dimension, Fig. 11.14 stage 4.

Release the workpiece and tilt in the vice at 30° with the aid of a protractor. Machine the angle, leaving the 32 mm dimension approximately 1 mm too long, to allow for machining the end face, Fig. 11.14 stage 5.

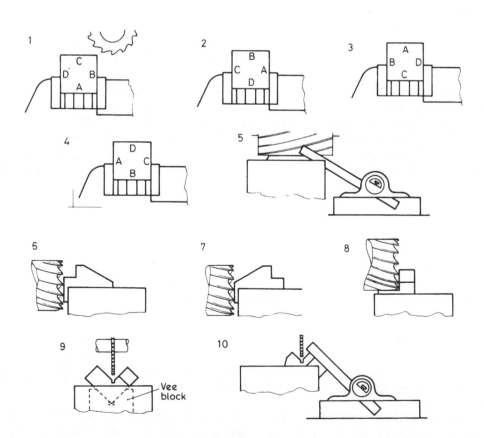

Figure 11.14 Sequence of operations on horizontal milling machine

Release the workpiece and reset with the end face protruding beyond the edge of the vice jaws, again seated on parallels. Mount a shell end mill on a stub arbor and load into the machine spindle.

Machine the end face to produce the 32 mm dimension to the start of the angle. At the same setting, machine the step to 20 mm and 40 mm dimensions, Fig. 11.14 stage 6.

Release the workpiece and reverse it, with the second end protruding beyond the edge of the vice jaws.

Machine to 80 mm length, Fig. 11.14 stage 7.

Release the workpiece and reset gripping on the 80 mm length.

Machine the step along the length to achieve the 25 mm and 28 mm dimensions, Fig. 11.14 stage 8.

Since the cutter on a horizontal machine cannot be inclined, milling the undercuts can be done only by tilting the workpiece.

Depending on the size and shape of the workpiece, setting can be simplified using a vee block. The vee block is set in the vice with the workpiece resting in the vee. The workpiece is then gripped in the vice across its ends, Fig. 11.14 stage 9.

Alternatively, a protractor is uscd to set the workpiece at 45°, Fig. 11.4 stage 10. These two final stages are carried out using a 5 mm side-and-face cutter mounted on a standard arbor.

Operations on a vertical machine

The stages of machining the same workpiece using a vertical machine are shown in Fig. 11.15. Although different cutters are used, the same basic principles apply.

Stages 1, 2, and 3 employ the same set up as for a horizontal machine, except that a shell end mill is used.

At stage 4, the step can be conveniently produced at the same setting and using the same cutter as for face D.

Stage 5 shows the angle being machined. The head of the vertical machine has been swivelled, so allowing the workpiece to be held in the vice seated on the parallels. This is a more convenient method if more than one workpiece is being machined. Each workpiece can be held in the vice in the normal manner, making it easier and quicker to set, with an assurance that the angle is identical on each item.

Stages 6 and 7 are set as before, but using an end mill to give a cutting edge long enough to machine the height of the workpiece and of large enough diameter to machine the step.

Machining the undercut at stages 8 and 9 again makes use of the swivelling head, with a 5 mm diameter end mill. Alternatively, with the head vertical the workpiece can be held for horizontal milling Stages 9 and 10.

Use of a rotary table

With the rotary table clamped to the worktable, the first essential is to bring the centre of the rotary table in line with the centre of the machine spindle. This is done by inserting a plug of the correct diameter into the hole in the centre of the rotary table.

Attach a dial indicator to the spindle nose and disengage the main gearbox to allow free rotation of the spindle. You may require a spanner on the drawbolt in order to rotate the spindle. Move the table and saddle traverses by hand to bring the plug

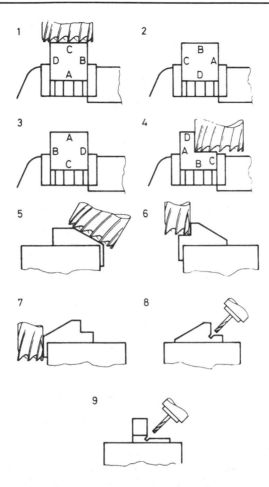

Figure 11.15 Sequence of operations on vertical milling machine

roughly central with the machine spindle. Rotate the spindle, and dial indicator attached to it, round the plug and adjust the table and saddle movements until a constant reading is obtained, Fig. 11.16. When this happens, the plug is central about the spindle centre line. Set the micrometer dial on each traverse movement to zero.

Having centred the rotary table, consider the workpiece shown in Fig. 11.17, which requires the 40 mm radius to be machined using an end-milling cutter in a vertical milling machine.

A bung is produced with one end to suit the hole in the rotary table and the other to suit the diameter of hole in the workpiece. Locate the bung in the rotary table and the workpiece on the bung. The workpiece is set on a pair of thin parallels, to raise the workpiece and avoid machining the table surface, Fig. 11.17.

Clamp the workpiece in position and mount the required size of end-mill cutter in the machine.

Lock the saddle movement in the already established central position. Move the machine table a distance equal to the radius to be machined plus half the cutter diameter. This distance is moved accurately by referring to the micrometer dial on the machine-table traverse.

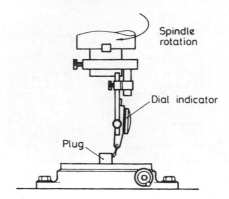

Figure 11.16 Centring a rotary table using a dial indicator

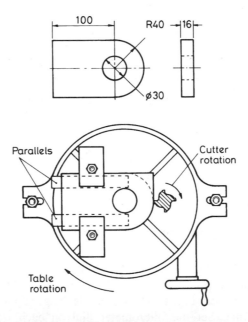

Figure 11.17 Workpiece set on rotáry table

Start the machine spindle, having selected the speed appropriate to the cutter size, and raise the knee to take a cut. Rotate the workpiece past the cutter until the complete radius is machined.

Raise the knee to take a further cut and repeat until all the surface is produced at the correct radius. Since a vertical milling machine is used, an alternative method of taking a cut can be used by lowering the spindle.

12 Joining methods

Some method of joining parts together is used throughout industry, to form either a complete product or an assembly. The method used depends on the application of the finished product and whether the parts have to be dismantled for maintenance or replacement during service.

There are five methods by which parts may be joined:

- mechanical fasteners – screws, bolts, nuts, rivets;
- soldering;
- brazing;
- welding;
- adhesive bonding.

Mechanical fasteners are most widely used in applications where the parts may need to be dismantled for repair or replacement. This type of joint is known as non-permanent. The exception would be the use of rivets, which have to be destroyed to dismantle the parts and so form a permanent joint. Welding and adhesives are used for permanent joints which do not need to be dismantled – any attempt to do so would result in damage to or destruction of the joints and parts.

Although soldered and brazed joints are considered permanent, they can be dismantled by heating for repair and replacement.

12.1 Mechanical fasteners

Machine screws

These are used for assembly into previously tapped holes and are manufactured in brass, steel, stainless steel, and plastics (usually nylon). Various head shapes are available, as shown in Fig. 12.1.

Countersunk   Raised countersunk   Round   Cheese   Raised cheese   Pan   Hexagon   Grub screw

Figure 12.1 Types of screw head

Depending on the style, thread diameters are generally available up to 10 mm, with lengths up to 50 mm. For light loading conditions where space is limited, a headless variety known as a grub screw is available. A typical application would be to retain a knob or collar on a shaft.

Socket screws

Manufactured in high-grade alloy steel with rolled threads, this type of screw is used for higher-strength applications than machine screws. Three head shapes are available, all of which contain a hexagon socket for tightening and loosening using a hexagon key, Fig. 12.2.

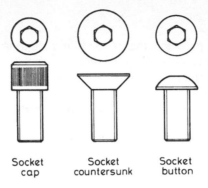

| Socket | Socket | Socket |
| cap | countersunk | button |

Figure 12.2 Socket screw heads

Headless screws of this type – known as socket set screws – are available with different shapes of point. These are used like grub screws, where space is limited, but for higher-strength applications. Different points are used either to bite into the metal surface to prevent loosening or, in the case of a dog point, to tighten without damage to the work, Fig. 12.3.

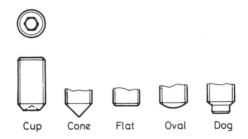

Cup   Cone   Flat   Oval   Dog

Figure 12.3 Socket set screws

Self-tapping screws

Self-tapping screws are used for fast-assembly work. They also offer good resistance to loosening through vibration. These screws are specially hardened and produce their own threads as they are screwed into a prepared pilot hole, thus eliminating the need for a separate tapping operation.

There are two types:

- the thread-forming type, which produces its mating thread by displacing the work material and is used on softer ductile materials, Fig. 12.4(a);
- the thread-cutting type, which produces its mating thread by cutting in the same way as a tap. This type has grooves or flutes to produce the cutting action, Fig. 12.4(b), and is used on hard brittle materials, especially where thin-wall sections exist, as this type produces less bursting force.

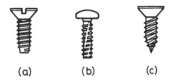

Figure 12.4 Self-tapping screws: (a) thread-forming, (b) thread-cutting, (c) self-piercing-and-tapping

More rapid assembly can be achieved by self-piercing-and-tapping screws. These have a special piercing point and a twin-start thread, Fig. 12.4(c). Used in conjunction with a special gun, they will pierce their own pilot hole in the sheet metal (up to 18 SWG (1.2 mm) steel) or other thin materials and are then screwed home in a single operation.

Bolts

Bolts are used in conjunction with a nut for heavier applications than screws. Unlike screws, bolts are threaded for only part of their length, usually twice the thread diameter.

Bright hexagon-head bolts are used in engineering up to 36 mm diameter by 150 mm long. Larger sizes are available in high-tensile materials for use in structural work.

Nuts

Standard hexagon nuts are used with bolts to fasten parts together. Where parts require to be removed frequently and hand tightness is sufficient, wing nuts are used, Fig. 12.5(a). If a decorative appearance is required, a dome or acorn nut can be fitted, Fig. 12.5(b).

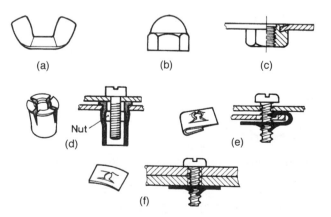

Figure 12.5 Types of nut: (a) wing, (b) dome or acorn, (c) rivet bush or nut, (d) blind, (e) flat spring steel fastner, and (f) J type fastner

Where thin sheets are to be joined and access is available from only one side, rivet bushes or rivet nuts are used. These provide an adequate length and strength of thread which is fixed and therefore allows ease of assembly, Fig. 12.5(c). Available in thread sizes up to 12 mm for lighter applications, blind nuts of the type shown in

Fig. 12.5(d) can also be used. The nut is enclosed in a plastics body which is pressed into a predrilled hole. A screw inserted into the nut pulls it up and, in so doing, expands and traps the plastics body.

Spring-steel fasteners are available which as well as holding also provide a locking action. If access is available from both sides, a flat nut can be used, Fig. 12.5(e), or from one side a J-type nut can be used, Fig. 12.5(f). In their natural state these nuts are arched, but they are pulled flat when the screw is tightened.

Washers

Washers distribute the tightening load over a wider area than does a bolt head, screw head, or nut. They also keep the surface of the work from being damaged by the fastener.

Plain flat washers spread the load and prevent damage, but do not provide a locking action. Washers which provide a locking action will be discussed in Section 12.3.

Spring tension pins

These pins are made from spring steel wrapped round to form a slotted tube, Fig. 12.6. The outside diameter is produced larger than the standard-size drilled hole into which it is to be inserted. When inserted in the hole, the spring tension ensures that the pin remains securely in position and cannot work loose. A chamfer at each end of the pin enables it to be easily inserted in the hole, where it can be driven home using a hammer.

Figure 12.6 Spring tension pin

Pins of this type are now being used to replace solid hinge pins, split pins, rivets, and screws, eliminating the need for reaming, tapping, counterboring, and countersinking. They are available in a range of diameters from 1 mm to 12 mm and lengths from 4 mm to 100 mm.

12.2 Screw threads

Since 1965, British industry has been urged to adopt the ISO (International Organisation for Standardisation) metric thread as a first-choice thread system, with the ISO inch (unified) thread as the second choice. The British Standard Whitworth (BSW), British Standard Fine (BSF), and British Association (BA) threads would then become obsolete. The British Standard Pipe (BSP) thread is to be retained. The changeover has been extremely slow in taking place, and all these threads are available and still in use.

ISO metric thread

This thread, based on a 60° triangular form, provides a range of coarse and a range of fine pitches (Appendix 1). The threads are designated by the letter M followed by the diameter and pitch in millimetres, e.g. M16 × 2.0. The absence of a pitch means that a coarse thread is specified, e.g. M16 indicates an M16 × 2.0 coarse pitch.

Unified thread

This thread is also based on a 60° form, but with a rounded crest and root. A coarse series of pitches (UNC) is provided from $\frac{1}{4}$ inch to 4 inch diameter, and a fine series of pitches (UNF) from $\frac{1}{4}$ inch to $1\frac{1}{2}$ inch diameter. The threads are designated by the diameter of thread in inches followed by the number of threads per inch and whether coarse or fine series, e.g. $\frac{1}{4}$–20 UNC, $\frac{1}{4}$–28 UNF (Appendix 2).

British Standard Whitworth (BSW) thread

This thread is based on a 55° vee-thread form, rounded at the crest and root, covering a range of thread diameters from $\frac{1}{8}$ inch to 6 inches. The threads are designated by the diameter in inches followed by the thread series, e.g. $\frac{3}{8}$ BSW (Appendix 3). It is not usual to include the number of threads per inch.

British Standard Fine (BSF) thread

This thread has exactly the same thread form as BSW but with finer pitches, covering a range from $\frac{3}{16}$ inch to $4\frac{1}{4}$ inches. The threads are designated by the diameter in inches followed by the thread series, e.g. $\frac{3}{8}$ BSF (Appendix 3).

British Standard Pipe (BSP)

This thread has exactly the same thread form as BSW and covers a range from $\frac{1}{8}$ inch to 6 inch diameter.

The BSP parallel threads, known as 'fastening' threads, are designated BSPF, the size referring to the bore size of the pipe on which the thread is cut, e.g. 1 inch BSPF has an outside diameter of 1.309 (Appendix 3).

Where pressure-tight joints are required, BSP taper threads have to be used. These have the same form and number of threads, but the thread is tapered at 1 in 16.

British Association (BA) thread

Threads in this range have extremely fine pitches and are used for applications less than $\frac{1}{4}$ inch diameter in preference to BSW or BSF. They are designated by a number, e.g. 4 BA (Appendix 4).

In general engineering, preference is given to the even-numbered BA sizes, i.e. 0, 2, 4, 6, 8, and 10 BA.

12.3 Locking devices

Self-locking screws and bolts

These screws and bolts eliminate the need for nuts and washers and so provide cost savings. Common types incorporate a nylon insert, either as a small plug or as a strip along the length of the thread, Fig. 12.7(a). A development of this principle is the application of a layer of nylon over a patch of thread. As the screw is engaged in the mating part, the nylon is compressed and completely fills the space between thread forms. This provides an interference which will resist rotation of the screw, Fig. 12.7(b). A more recent development is the application to an area of thread of a chemical adhesive which is completely dry to the touch, Fig. 12.7(c). The liquid adhesive is encapsulated within the film. When the two threaded parts are assembled,

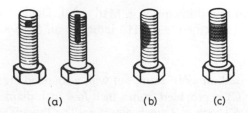

Figure 12.7 Self-locking screws and bolts: (a) nylon insert, (b) nylon layer, (c) adhesive film

the micro-capsules of adhesive are broken, releasing the adhesive, which hardens and provides a reliably sealed and locked thread.

Locking nuts

The simplest method of locking a nut in position is by applying a lock nut. Lock nuts are a little over half the thickness of a standard nut. When used in conjunction with a standard nut and tightened, the lock nut is pushed against the thread flanks and locked, Fig. 12.8(a).

Slotted and castle nuts are used in conjunction with wire or a split pin through a hole in the bolt to prevent the nut from working loose, Fig. 12.8(b).

Self-locking nuts are available which are easy to assemble and do not require a hole in the bolt or the use of a split pin. One type, known as a 'Nyloc' nut, Fig. 12.8(c), incorporates a nylon insert round the inner top end of the nut. As the nut is screwed on, the nylon yields and forms a thread, creating high friction and resistance to loosening.

A second type, known as an 'Aerotight' stiff nut, Fig. 12.8(d), has two arms formed on top of the nut. These arms, which are threaded, are deflected inwards and downwards. When the nut is screwed on, the arms are forced into their original position and the resistance of these arms gives a good grip on the thread, preventing it from working loose.

A third type, known as a 'Philidas' self-locking nut, Fig. 12.8(e), has a reduced diameter above the hexagon. Two slots are cut opposite each other in the reduced diameter and the metal above the slots is pushed down, which upsets the thread pitch. When screwed in position, the thread is gripped by the upset portion, preventing the nut from working loose.

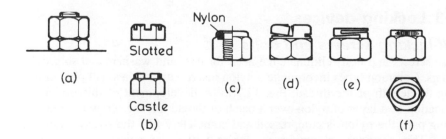

Figure 12.8 Locking nuts: (a) standard lock nut, (b) slotted and castle nuts, (c) 'Nyloc' nut, (d) 'Aerotight' nut, (e) 'Philidas' self locking nut, (f) torque lock nut

A fourth type, known as a torque lock nut, Fig. 12.8(f), has the top part of the nut deformed to an elliptical shape which grips the thread as the nut is applied. This ensures close contact between the threads, preventing the nut from working loose.

Locking washers

A locking washer is inserted under the head of a screw, bolt, or nut to prevent it working loose during service.

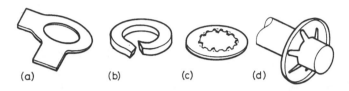

Figure 12.9 Locking washers: (a) tab, (b) helical spring, (c) shake proof, (d) spring fixing

A tab washer may be used, similar to a plain washer with the addition of a tab which is bent up on the hexagon face of the nut, screw, or bolt to prevent it working loose, Fig. 12.9(a).

Helical-spring locking washers are commonly used as locking devices and are available for threads up to 24 mm diameter. They may be of square or rectangular section in a single coil, with the ends of the coil raised in opposite directions. These ends form sharp points which dig into the surfaces. In addition, the spring is flattened during the tightening of the screw, bolt, or nut, which gives constant tension during use, Fig. 12.9(b).

Shake-proof washers are used for thread sizes up to 16 mm and can have external or internal teeth. The teeth are twisted out of flat so that the washer bites into the surfaces as it is compressed during tightening, Fig. 12.9(c).

Where rigid permanent fixing is required on shafts, a range of spring fixing washers which eliminate the use of threads and nuts is available. One type is shown in Fig. 12.9(d). As it is pushed on to the shaft, the 'prongs' are deformed and bite into the shaft and cannot be removed without destroying it. This type is available up to 25 mm diameter and can be used on all types of material, including plastics.

12.4 Riveting

Solid and tubular rivets

Riveting as a means of fastening is used because of its speed, simplicity, dependability, and low cost. Light riveting, used for general assembly work up to about 6 mm diameter, is carried out in industry using high-speed rivet-setting machines having cycle times as short as $\frac{1}{3}$ second. Rivets are used on assemblies where parts do not normally have to be dismantled, i.e. permanent joints. They may also be used as pivots, electrical contacts and connectors, spacers, or supports.

The cost of riveting is lower than that of most other methods of fastening, due to the absence of plain washers, locking washers, nuts, or split pins; also, the use of self-piercing rivets eliminates the need to predrill holes. Rivets are available in steel,

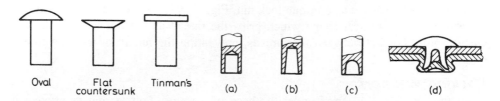

Figure 12.10 Solid-rivet head types

Figure 12.11 Tubular rivets: (a) short-hole, (b) double-taper semi-tubular, (c) bull-nose semi-tubular, (d) self-piercing

brass, copper, and aluminium in a variety of types. The more standard head types used are shown in Fig. 12.10.

Solid rivets are strong but require high forces to form the end. In riveting, forming the end is known as clinching. Solid rivets are used in applications where the high forces used in clinching will not damage the work being fastened. Tubular rivets are designed for application where lighter clinching forces are used.

Short-hole tubular rivets, Fig. 12.11(a), have the advantage of a solid rivet with easier clinching. These have a parallel hole and can be used for components of varying thicknesses.

Double-taper semi-tubular rivets, Fig. 12.11(b), have a taper hole to minimise shank expansion during clinching and are used to join brittle materials.

Bull-nose semi-tubular rivets, Fig. 12.11(c), are used where maximum strength is required. The rivet shank is intended to expand during the setting operation in order to fill the predrilled hole in the work. This ensures a very strong joint.

Self-piercing rivets, Fig. 12.11(d), have been specially developed to pierce thicker metal and clinch in the same operation. For metal up to 4.7 mm thick, the self-piercing rivet is made from special steel and is heat-treated to give the required hardness for piercing and ductility for clinching. One use of this type of rivet is automatic riveting in the production of garage doors.

Blind rivets

Blind rivets, also known as pop rivets, are rivets which can be set when access is limited to one side of the assembly. However, they are also widely used where both sides of the assembly are accessible.

Used to join sheet metal, blind rivets are readily available in sizes up to around 5 mm diameter and 12 mm long in aluminium, steel and monel. Plated steel rivets are used where low cost, relatively high strength, and no special corrosion-resistance is required; aluminium for greater resistance to atmospheric and chemical corrosion; and monel for high strength and high resistance to corrosion.

Blind rivets consist of a headed hollow body inside of which is assembled a centre pin or mandrel. The rivet is set by inserting the mandrel in a tool having a means of gripping, and the rivet is inserted into a predrilled hole in the assembly. Operation of the tool causes its gripping jaws to draw the mandrel into the rivet, with the result that the head of the mandrel forms a head on the rivet on the blind side of the assembly, at the same time pulling the metal sheets together. When the joint is tight, the mandrel breaks at a predetermined load, Fig. 12.12. The broken-off portion of the mandrel is then ejected from the tool.

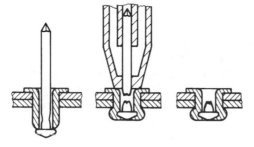

Figure 12.12 Blind rivet

12.5 Soft soldering

Soft soldering is a process of joining metal parts by heating and running a low-melting-point alloy between the two surfaces being joined. When cooling takes place, the alloy solidifies, resulting in a secure joint. For most purposes, the low-melting-point solder alloy is a composition of tin and lead. Solder melts at a temperature less than 300°C, which is far below the melting temperature of the metal being joined, and produces a low-strength joint.

Soldering is used not only to make a mechanical bond between surfaces but also to provide a leak-proof seal when liquids have to be contained. It can also be used to provide a permanent electrical connection. To provide a secure joint, it is essential that the joint surfaces are perfectly clean and free from rust, grease, or any other substance likely to prevent good metal-to-metal union. The most likely cause of a bad joint, assuming the surface is clean, is the thin oxide film which is present on all metals. Oxide films can be removed using emery cloth or a flux.

Fluxes

Fluxes used in soft soldering are either active or passive.

An active flux chemically removes the oxide film, has an acid base, and is highly corrosive. These fluxes are usually hydrochloric acid in which zinc has been dissolved to form zinc chloride, known as 'killed spirits'. Any joint prepared using an active flux must be thoroughly washed in warm water when soldering is completed, to remove any flux residue. For this reason an active flux is not suitable for electrical applications.

A passive flux is used after the oxide film has been removed using emery cloth, to prevent the oxide film reforming. Passive fluxes are usually resin-based. Cored solder, containing a resin-based flux, is usually sufficient for electrical work.

Heating

The type of heat source used for soldering depends largely on the size of the parts to be joined. The greatest problem is usually heat loss to the surrounding area by conduction through the metal. The temperature in the joint area must be high enough to melt the solder and allow it to flow and combine with the surfaces to be joined.

Where the conduction of heat is likely to cause damage, e.g. to electrical insulation or electronic components, a piece of bent copper can be placed in contact with the conductor, positioned between the joint and the insulation or component. The copper readily absorbs the heat before it can do any damage and is known as a 'heat sink'.

For small parts, an electric soldering iron is quite sufficient. The soldering iron, as well as melting the solder and heating the work, acts as a reservoir for the solder to deposit an even amount in the required position. Larger parts can be heated using a gas/air or butane torch or by placing the work on a hot-plate.

When the area of the surfaces being joined is large, solder cannot be satisfactorily run between them. In such a case, the surfaces should first be 'tinned', i.e. each surface should be separately coated with solder. The two parts are then assembled and the parts are reheated until the solder melts to make the joint.

Joint design

Solder is not as strong as other metals. When a mechanical joint is required, the joint design should provide as much additional strength as possible and not rely on the solder strength alone. A higher mechanical strength can be obtained by an interlocking joint, the solder providing an additional leak-proof seal as shown in Fig. 12.13(a). This method is used in the production of cans of drink and food stuffs.

Additional mechanical strength can be obtained in electrical joints by winding the wire round a pillar or bending the wire through a hole in a tab connector, as shown in Fig. 12.13(b) and (c).

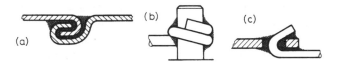

Figure 12.13 Soldering-joint design: (a) interlocking, (b) winding round a pillar, (c) bending through a tab connector

12.6 Solders

Soft solders are alloys of tin and lead. All the plain tin/lead solders become solid at 183°C. The temperature at which they become completely liquid depends on the composition, the temperature increasing as the lead content increases.

Some solders pass through a considerable pasty stage from being completely liquid to becoming solid. Reference to Fig. 12.14 shows a 20% tin/80% lead solder, completely liquid at 276°C and solid at 183°C. The solder passes through a pasty stage in the transformation from liquid to solid. This feature is useful in some plumbing applications or where work is coated by dipping in a bath of molten solder.

A solder which contains 60% tin/40% lead has a lower melting temperature and a very small temperature interval between being completely liquid and becoming solid, 188°C to 183°C. This is preferred, especially in electrical work, where a higher melting temperature and longer cooling period could result in damage to insulation or components. Where soldering is carried out by machine and too fine a control of temperature cannot be maintained, the 50/50, 40/60 ranges of solders are used.

Table 12.1 shows the temperatures, strength, and uses of a range of compositions of plain tin/lead solder alloys.

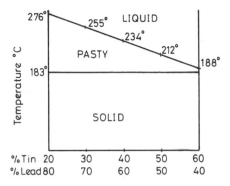

Figure 12.14 Tin–lead diagram

Table 12.1 Tin/lead solders

% tin	% lead	Solidus* (°C)	Liquidus† (°C)	Strength (MN/m²)	Uses
60	40	183	188	58	General purpose, especially electrical works
50	50	183	212	46	} Machine solders
40	60	183	234	41	
30	70	183	255	38	} Plumbers' solders and
20	80	183	276	37	} dipping baths

* Solidus – the temperature at which the alloy has completely solidified.
† Liquids – the temperature at which the alloy is completely liquid.

12.7 Brazing

Brazing is defined as a process of joining metals in which, during or after heating, molten filler metal is drawn by capillary action into the space between closely adjacent surfaces of the parts being joined. In general, the melting point of the filler metal is above 450°C but always below the melting temperature of the metals being joined.

Brazing is used to join any combination of similar or dissimilar metals and results in a high-strength joint of good reliability.

To form a strong joint, the surfaces must be free of any rust, grease, or oxide film.

Brazing alloys

Brazing alloys are available in a wide variety of forms, including rod, strip, wire, foil, and powder.

The choice of brazing alloy depends upon the materials being joined and the temperature at which the brazed parts are to operate. Brazing brasses are widely used with hand-torch heating for joining ferrous-metal parts. The common composition of brazing brass is shown in Table 12.2.

Silver brazing alloys have excellent brazing properties and are the most widely used for joining most ferrous and non-ferrous materials with the exception of those based on aluminium, zinc, and magnesium. Silver brazing alloys have lower melting points than the brazing brasses and are capable of penetrating narrow joint gaps. The composition of two typical silver brazing alloys is shown in Table 12.3.

Table 12.2 Brazing brass

Composition %		Melting range (°C)	
Copper	Zinc	Solidus	Liquidus
60	40	885	890

Table 12.3 Silver brazing alloys

Composition %				Melting range (°C)	
Silver	Copper	Zinc	Cadmium	Solidus	Liquidus
42	17	16	25	610	620
44	30	26	–	675	735

Cadmium has the effect of lowering the melting temperature but can produce toxic fumes during brazing. These fumes must be extracted by a ventilating system to avoid any risk of inhalation. If ventilation is not possible, a cadmium-free alloy should be used.

Aluminium-silicon alloys are used for the brazing of aluminium and aluminium alloys. The composition of a commonly used aluminium brazing alloy is shown in Table 12.4.

Table 12.4 Aluminium brazing alloy

Composition %			Melting range (°C)	
Aluminium	Silicon	Copper	Solidus	Liquidus
86	10	4	535	595

Fluxes

The function of a flux is to dissolve or remove the surface oxide film from the metal to be joined and any oxides formed during heating. The ideal flux should be active at a temperature below the solidus and remain active at a temperature above the liquidus. No single flux can achieve this over the range of temperatures used in brazing and so a flux has to be chosen to suit a particular temperature range.

Borax is a cheap and readily available flux and is widely used for general applications. It is supplied as a powder and mixed with water to form a paste. Ideally the flux should be applied as a paste before heating. Borax/boric-acid mixtures with other additives are used with the silver-base brazing alloys.

These fluxes must be removed after brazing. This can be done by quenching the work in hot or cold water shortly after the brazing alloy has solidified. If this is not practical, the flux can be removed by chipping, filing, scraping, or steel-wire brushing.

Heating

Any heat source capable of raising the temperature above the liquidus of the selected brazing alloy can be used. Many types of controlled automatic heating are used in industry, but for workshop purposes the hand torch is the most widely used. A hand torch has the advantage of being flexible in use, but it requires a skilled operator to produce consistent results.

A wide variety of gas mixtures can be used, the most common being

- oxy–acetylene,
- oxy–propane,
- compressed air + coal gas,
- compressed air + natural gas.

When using a hand torch, care must be taken to achieve an even distribution of heat, especially when using oxy–acetylene with its intensely hot localised flame.

Joint design

The strength of a brazed joint relies on the capillary action of the brazing alloy between the faces being joined.

Wherever possible, joints should be designed so that loads applied in service act on the joint as shear stresses rather than as tensile stresses. This means that lap joints are preferred to butt joints, Fig. 12.15. The recommended length of overlap on lap joints is between three and four times the thickness of the thinnest component in the assembly.

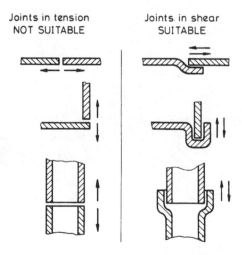

Figure 12.15 Brazing – joint design

To take full advantage of the capillary action, it is essential that there is a sufficient gap or clearance between the faces being joined, to allow penetration of the brazing alloy. Depending on the metals being joined, the joint gap should be between 0.04 mm and 0.20 mm.

12.8 Welding

Welding differs from soldering and brazing in that no alloy is used to join the metals. The metals being joined are locally melted and when solidified produce a solid mass. The joint strength is therefore as strong as the metals being joined. A filler rod, with a melting temperature similar to that of the metals being joined, is sometimes used to make up losses during welding, to fill any gap between the joint surfaces, and to produce a fillet. A flux is required with some metals and some welding methods, to remove the oxide film and provide a shield to prevent oxides reforming.

In order to melt metals to be joined by fusion, a source of high heat energy is required. The two chief methods of obtaining the high temperature required are by an electric arc and by burning a mixture of gases.

Arc welding

An electric arc is produced by passing an electric current between two electrodes separated by a small gap. In arc welding, one electrode is the welding rod or wire, the other is the metal plate being joined.

The electrodes are connected to the electrical supply, one to the positive terminal and one to the negative. The arc is started by touching them and withdrawing the welding rod about 3 or 4 mm from the plate. When the two electrodes touch, a current flows, and, as they are withdrawn, the current continues to flow in the form of a spark. The resulting high temperature is sufficient to melt the metal being joined. The circuit is shown in Fig. 12.16(a).

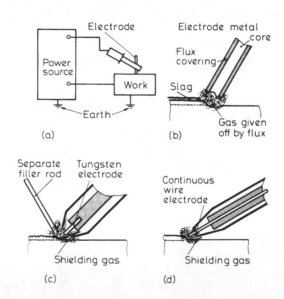

Figure 12.16 Arc welding: (a) circuit, (b) MMA Welding, (c) TAGS welding, (d) MAGS welding

When the electrode also melts and deposits metal on the work, it is said to be consumable. Electrodes made from tungsten which conduct current but do not melt are known as non-consumable.

The most common arc-welding methods are manual metal arc welding, tungsten arc gas-shielded welding, and metal arc gas-shielded welding.

Manual metal arc (MMA) welding, Fig. 12.16(b): In this process the arc is struck between a flux-covered consumable electrode and the work. This method is the most widely used form of arc welding and is used on all materials with the exception of aluminium. The flux produces gas which shields the surface of the molten metal and leaves behind a slag which protects the hot metal from the atmosphere while cooling and has to be chipped off when cool.

Tungsten arc gas-shielded (TAGS) welding, Fig. 12.16(c): In this process the arc is struck between a non-consumable tungsten electrode and the workpiece. The tungsten electrode is held in a special gun through which argon gas flows to shield the electrode and molten metal from atmospheric contamination – the process is often referred to as TIG or argon arc welding. Additional filler metal can be applied separately as rod or wire. The argon shield enables aluminium, magnesium alloys, and a wide range of ferrous metals to be welded without the use of a flux. This method is used primarily for welding sheet metal and small parts and produces a high-quality weld.

Metal arc gas-shielded (MAGS) welding, Fig. 12.16(d): In this process the arc is struck between a continuous consumable wire electrode fed through a special gun. A shielding gas – argon, carbon dioxide (CO_2), oxygen, or a mixture of these – is also fed through the gun to shield the arc and molten metal from contamination. Using different filler wires and types of gas, this method is suitable for welding aluminium, magnesium alloys, plain-carbon and low-alloy steels, stainless and heat-resisting steels, copper, and bronze.

Using carbon-dioxide shielding gas for plain-carbon and low-alloy steels, this method is referred to as CO_2 welding.

Gas welding

The heat for this method is obtained by burning a mixture of a combustible gas and oxygen to provide an intense flame. The two gases are mixed in the correct proportion in the mixing chamber of the welding torch. Control of the gas mixture is by two valves in the handle of the torch. Interchangeable nozzles of different sizes are fitted to the torch to give a range of flame intensities to suit a variety of applications. The most widely used combustible gas is acetylene, which when mixed with oxygen gives a flame temperature of around 3250°C.

The flame is directed on to the metal being joined, resulting in a molten pool of metal. Filler metal is added when required and is usually the same material as that being joined. Fluxes may be required, as in the case of welding aluminium or copper alloys.

This method is used to weld a wide variety of metals. It is cheap and portable, making it invaluable for repair work. Although it can be used on steel up to 25 mm thick, it is generally employed on sheet material 16 SWG (1.6 mm) and below. Arc welding is used on sheet material above this thickness, as less distortion is produced, due to the more localised heat source of the arc.

12.9 Adhesives

Adhesives are used to establish a permanent bond between two joint surfaces. Adhesive bonding is the modern term for glueing, and the technique is used to join metals to themselves and also a wide variety of metallic and non-metallic materials.

Many types of adhesive are available, their applications varying from use in surgery to heal internal wounds and organs to joining structural members in aircraft.

The bonding process basically consists of thoroughly cleaning the surfaces to be joined, preparing and applying the adhesive, and finally assembling the parts. The complete process may be carried out at room temperature or at a higher temperature, with or without pressure and for a period depending on the time required for the adhesive to set (or cure).

There are advantages and disadvantages in using adhesives, some of which are listed below.

Advantages

- A variety of materials can be bonded. These may be similar or dissimilar; thick or thin; metallic or non-metallic.
- Thin, delicate, and heat-sensitive parts which heat methods of joining would distort or destroy can be bonded.
- When used to replace mechanical methods of joining:

    i) hole drilling is eliminated – reducing time,
                                      – reducing cost,
                                      – avoiding weakness in the region of the hole;
    ii) a weight saving is made;
    iii) the outer surface is smooth and free from bolt and rivet heads.
- The adhesive layer provides:
    i) a good seal against moisture,
    ii) good thermal and electrical insulation,
    iii) some flexibility of the joint.

Disadvantages

- Problems may exist:

    i) in preparing joint surfaces,
    ii) in storing, preparing, and applying the adhesive,
    iii) in the time required for curing.
- Bonded structures are difficult to dismantle for repair and replacement.
- In most cases, temperature limitations in service are below those of other joining methods.
- Health and fire hazards when using solvent-based fluids.

Types of adhesive

The types of adhesive in common use are shown in Fig. 12.17.

Adhesives can be conveniently grouped under two headings: natural and synthetic.

Natural adhesives

These can be subdivided into adhesives of animal, vegetable, and mineral origin. Most natural adhesives produce low-strength joints with poor resistance to water and

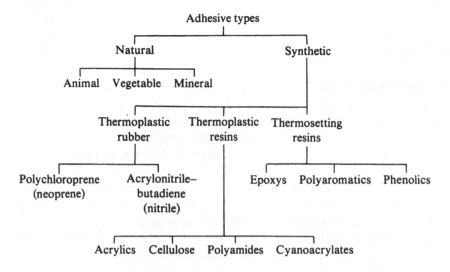

Figure 12.17 Types of adhesive

to temperature variation. The exception is the mineral type, which has excellent high-temperature properties.

Applications of natural adhesives are mainly in woodworking and packaging industries.

Synthetic adhesives

These can be subdivided into thermoplastic resins, thermoplastic rubbers, and thermosetting resins.

Thermoplastic resins: These soften when heated and do not change chemically when establishing a bond.

Their application is confined to low-load assemblies formed from metals, ceramics, glass, plastics, and porous materials based on paper, wood, leather, and fabrics which are not subject to severe service conditions.

These adhesives are available in liquid and solid form. Solid forms are available as a film which requires heat to convert it to a liquid state. Bonding is usually carried out under heat and pressure after joint assembly. One of the most common types is:

Cyanoacrylate (superglue) This is a one-component liquid adhesive which establishes a bond at room temperature without the addition of a catalyst. The adhesive reacts with the small amount of moisture present on most surfaces. The joint surfaces are cleaned and a few drops of adhesive are applied. The two parts are brought together quickly under pressure and are held for a few minutes. Depending on conditions, the bond may be complete in as short a time as 15 seconds.

This adhesive is non-toxic and is used in surgery for healing internal wounds and organs. Industrial uses are for light structures requiring fast assembly, such as instrument components and electronic and optical units.

This is an expensive adhesive but has the advantage of eliminating the need for expensive heating and pressure equipment. Instant adhesion to the skin is a major hazard. If the human skin is accidentally bonded, it can be separated by a gentle peeling action aided by a blunt instrument and washing with a detergent.

It is available under the brand name 'Loctite Superfast'.

Thermoplastic rubbers: These are some of the most versatile adhesives currently in use. Most are available as solvent forms, i.e. the solvent evaporates and establishes a bond. As with the thermoplastic resins, these adhesives soften when heated and do not change chemically when the bond is established. Application is confined to low-load assemblies. One of the most common types is:

Polychloroprene rubbers (neoprene) These are solvent-based liquid adhesives compounded with fillers to give additional properties. The adhesive sets due to solvent evaporation, which can be accelerated by heat.

They have good resistance to water, salt spray, chemicals, lubricants, and weak acids, but are unsuitable for structural applications. They have a limited service range, from −50°C to 95°C, and are used as a general-purpose contact adhesive for metals, plastics laminates, decorative wall panels, fabrics, rubbers, and interior car trim.

They are available under the brand names 'Bostik' and 'Evo-stik' impact.

Thermosetting resins: These adhesives change chemically when the bond is established under the action of heat, a catalyst, or both. They provide a range of structural adhesives for high-load applications and exposure to severe environmental conditions such as heat, cold, humidity, and chemical atmospheres.

Thermosetting adhesives are supplied as liquids, pastes, and solids. Liquids are generally two-component systems which, when mixed, cure by catalytic action with or without the aid of heat. Pastes can be applied to vertical joints and do not flow out during assembly and cure.

Film is available in various thicknesses and has the advantages of easy clean handling and that it can be cut to the joint shape. Curing with heat and pressure is necessary. One of the most common types is:

Epoxys These are thermosetting adhesives in which the bond is obtained from the reaction of a resin and a basic curing agent (hardener). In general, these are two-component adhesives which are mixed just before use, applied within a certain time, and cured at room temperature or at an elevated temperature to reduce the cure time.

They provide excellent bond strength with good resistance to weathering, contact with oils, greases, fuels, acids, alcohols, heat, and cold. They are used as a structural adhesive for many materials: metal, glass, ceramics, wood, concrete, and thermosetting plastics; also

- for bonding aluminium to itself and other materials in aircraft,
- for bonding reinforcements to light-gauge panels for greater rigidity in the motor industry,
- for bonding laminated beams in woodworking, and
- for bridge and road repairs in the building industry.

They are available under the brand name 'Araldite'.

12.10 Electrical connections

Any electrical connection must securely anchor all the wires of the conductor and not place any appreciable mechanical stress on the terminal.

Mechanical connections

The simplest form of mechanical connection is where the wire is looped round clear of the diameter of a screw and is firmly clamped by a washer under the screw head or, alternatively, by means of a washer and nut, Fig. 12.18(a) and (b). Another form is the brass pillar terminal found in plugs, sockets, and lampholders, which has a hole drilled through it in which the wire is securely held by means of a brass screw, Fig. 12.18(c).

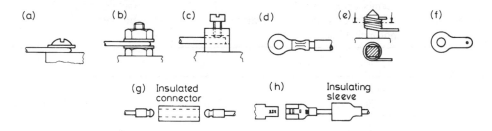

Figure 12.18 Electrical connections: (a) washer, (b) nut, (c) pillar terminal, (d) socket, (e) wrapped joint, (f) tag, (g) bullet connector, (h) quick-connect connector

    Connection to a socket can be made by squeezing or crimping directly to the wire. Alternatively, the wire can be soldered into the socket, Fig. 12.18(d).

    Wrapped joints are used to connect wires to terminal posts, Fig. 12.18(e). The terminal post, called a wrapping post, is square or rectangular in section. Several turns of the wire are twisted under pressure round the post, using a special wrapping tool. The electrical connection is made by the wrapped wire digging into the corners of the post.

Soldered connections

Solder alone should not be relied upon to make a secure connection. The wire should be bent or wrapped round to give good mechanical strength and then be soldered to give the required electrical connection. Sufficient solder should be applied to enable the wire to be seen through the solder, and with just enough heat to allow the solder to flow freely round the connection (see Section 12.5).

    Soldered connections are made directly to a circuit or to tags of various design, one of which is shown in Fig. 12.18(f).

Vehicle connections

Electrical connections on vehicles may need to be easily disconnected for replacement of components. One such type is the 'bullet' connector, shown in Fig. 12.18(g), where the ends are soldered or crimped to the wire and then pushed firmly into the insulated connector. Quick-connect types as shown in Fig. 12.18(h), usually crimped to the wire, are pushed together to form the connection. An insulating sleeve can then be pushed over the completed connection.

12.11 Relative merits of joining methods

	Heat requirement	Type of joint	Heat conductivity
Soldering	Less than 300°C	Permanent – can be dismantled using heat	Not above 183°C
Brazing	Above 450°C	Permanent – can be dismantled using heat	Not above the solidus of the brazing alloy used
Welding	Melting point of metals being joined	Permanent	Up to melting point of metals joined
Adhesives	None (except the hot-melt films)	Permanent	Insulator

	Electrical conductivity	Strength of joint	Type of material joined
Soldering	Conductor	Low	Similar and dissimilar metals
Brazing	Conductor	Medium	Similar and dissimilar metals
Welding	Conductor	High	Similar metals
Adhesives	Insulator	Low to high	Similar and dissimilar metals and non-metals

13 Materials

A wide range of materials is used in engineering, and it is important to be aware of the ways in which these are applied and of the properties which make them suitable for these applications.

Properties of materials can be divided into two groups: physical and mechanical. Physical properties are those properties of a material which do not require the material to be deformed or destroyed in order to determine the value of the property. Mechanical properties indicate a material's reaction to the application of forces. These properties require deformation or destruction tests in order to determine their value. The value of these properties can be altered by subjecting the material to heat treatment and cold or hot working.

13.1 Physical properties

Coefficient of linear expansion

This is a measure of the amount by which the length of a material increases when the material is heated through a one-degree rise in temperature. Thus

$$\begin{array}{c} \text{increase} \\ \text{in length} \end{array} = \begin{array}{c} \text{original} \\ \text{length} \end{array} \times \begin{array}{c} \text{temperature} \\ \text{rise} \end{array} \times \begin{array}{c} \text{coefficient of} \\ \text{linear expansion} \end{array}$$

Thus, if the coefficient of linear expansion of copper is 0.000017 per °C (written 0.000017/°C or 17×10^{-6}/°C) then for each degree rise in temperature a length of copper will expand by 0.000017 of its original length; for example, a 100 mm long copper rod will expand $0.000017 \times 100 = 0.0017$ mm for each degree rise in temperature. If this 100 mm long copper rod is heated through 20°C, then the amount of expansion will be $100 \times 0.000017 \times 20 = 0.034$ mm.

Different metals expand or contract by different amounts for a given temperature change; for example, aluminium expands at a greater rate than cast iron. That different metals have different values for the coefficient of linear expansion can be useful on some occasions while on other occasions it can be a disadvantage.

A typical application of advantage is in the construction of a thermostat. This device makes use of two strips of different materials clamped together, the different expansion rates when heated causing the strip to bend and so make or break an electrical contact, Fig. 13.1.

The disadvantages are many and have to be allowed for during design. For instance, the clearance between the aluminium piston and the cast-iron cylinder block in motor-vehicle engines will be less when the engine is hot than when it is cold.

Specific heat capacity

The specific heat capacity of a material is the amount of heat energy (in joules) required to raise the temperature of unit mass (one kilogram) of the material by unit rise in temperature (one degree Celsius).

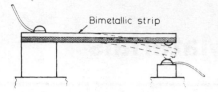

Figure 13.1 Bimetallic strip thermostat

The quantity of heat energy required to raise the temperature of a piece of material depends upon the type of material and its mass. Thus equal masses of two different materials will require different amounts of heat energy to raise their temperature by the same amount; for example, 1kg of water with a specific heat capacity of 4200J/(kg °C) will require 4200 joules of heat energy to raise its temperature by one degree Celsius. Similarly, 1kg of copper of specific heat capacity 386J/(kg °C) will require 386 joules to raise its temperature by one degree Celsius.

A knowledge of this property is required when dealing with heating or cooling operations. Various liquids are used when cooling after heat-treatment operations. The liquid used must be able to absorb the heat energy from the block of metal, and this depends upon the type of liquid and its mass. In metal-cutting operations, the coolant used must be of a type and delivered in sufficient volume to remove heat from the tool and cutting area without itself becoming too hot.

Density

Equal volumes of different materials have different masses. The mass in a given volume is a measure of the density.

The density of a material is the mass per unit volume and is given by the equation

$$\text{density} = \frac{\text{mass of material}}{\text{volume occupied by the material}}$$

Since mass is measured in kilograms and volume in cubic metres, the unit of density is kilograms per cubic metre, written kg/m^3. For example, the density of aluminium is 2700 kg/m^3 and that of lead is 11300 kg/m^3. This means that a volume of one cubic metre of aluminium has a mass of 2700 kg and the same volume of lead has a mass of 11300 kg.

This property must be considered where the mass has to be restricted. In the production of aircraft, for instance, special materials have been developed which are as strong as steel but are only a fraction of its density.

Melting point

This is the temperature at which a material changes from the solid to the liquid state. This may be an important consideration in some material applications; for example, it is important to know the melting point of a solder if it is used on a joint which may be subjected to temperatures approaching the solder's melting point. The many plastics now available must be used within their temperature limits. Equipment used in hot-working processes – such as furnaces, casting machines, and forging dies – must be designed to withstand their high working temperatures.

Thermal conductivity

When one end of a metal bar is heated, the heat will be conducted along the length of the bar. The rate at which the heat is conducted depends on the bar material, some materials being better conductors than others. This heat-conducting ability of a metal is measured by the thermal conductivity.

Thermal conductivity is given as a rate of transfer of heat energy, measured in J/(m s °C) – that is, the number of joules of energy transferred per second through one metre for each degree Celsius rise of temperature. Since one joule per second equals one watt, thermal conductivity can also be quoted in W/(m °C).

A good conductor such as copper has a high thermal conductivity and is used where heat has to be readily transferred, e.g. in a soldering iron or a motor-car radiator. Bad conductors, such as the non-metallic materials, have a low thermal conductivity and are used where heat has to be retained, e.g. lagging materials on hot-water tanks and pipes.

Electrical resistivity

Some materials allow electricity to pass through them very easily and are electrical conductors. These include carbon and most of the metals, such as aluminium, copper, brass, and silver. Other materials offer a high resistance to the flow of electricity and are bad electrical conductors, known as insulators – these include non-metallic materials such as plastics, rubber, mica, ceramics, and glass.

The resistance of an electrical conductor is measured in ohms and depends on the dimensions of the conductor as well as the material from which it is made. It is fairly easy to see that conductors of similar shape but made of different materials may have different resistances.

In order to compare the resistance effect of different conductor materials, a standard size and shape of conductor is considered. The standard shape chosen is a cube whose sides are one metre. The resistance of this metre cube of material is known as the resistivity of the material, measured in ohm metres. Thus an electrical-conducting material will have a low resistivity while an insulator will have a high resistivity.

The resistance to the flow of electricity can be found from the following equation, knowing the area and length of the conductor and its resistivity:

$$R = \frac{\rho l}{a}$$

where R = resistance, in ohms

l = length of conductor, in metres

a = cross-sectional area of conductor, in square metres

ρ = resistivity of the conductor material, in ohm metres

❏ Example 13.1

What is the resistance of an electrical conductor 1 mm diameter and 20 metres long whose resistivity is 2.5×10^{-8} ohm metres?

$$\text{area of conductor} = \frac{\pi \times 1^2}{4} = 0.7854 \text{ mm}^2$$

Since the other values are expressed in metres,

$$\text{area in m}^2 = \frac{0.7854}{1000 \times 1000} = 0.7854 \times 10^{-6}\text{m}^2$$

Resistance is given by $\rho l/a$ which is

$$\frac{2.5 \times 10^{-8} \times 20}{0.7854 \times 10^{-6}} = 0.637 \text{ ohms}$$

13.2 Mechanical properties

Hardness

A material which is hard is able to resist wear, scratching, indentation, and machining. Its hardness is also a measure of its ability to cut other materials. Hard materials are required for cutting tools and for parts where wear must be kept to a minimum.

Brittleness

A brittle material will break easily when given a sudden blow. This property is associated with hardness, since hard materials will often be brittle. Brittle materials cannot be used in the working parts of power presses, which are subjected to sudden blows.

Strength

A strong material is able to withstand loads without breaking. Loads may be applied in tension, compression, or shear, and a material's resistance to these loads is a measure of its tensile strength, compressive strength, and shear strength. Connecting rods in internal-combustion engines must be strong in tension and compression, while the gudgeon pin must be strong in shear, Fig. 13.2.

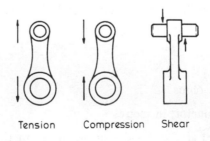

Tension   Compression   Shear

Figure 13.2 Tension, compression and shear

Ductility

A ductile material can be reduced in cross-section without breaking. In wire-drawing, for instance, the material is reduced in diameter by pulling it through a circular die. The material must be capable of flowing through the reduced diameter of the die and at the same time withstand the pulling force.

Malleability

A malleable material can be rolled or hammered permanently into a different shape without fracturing. This property is required when forging, where the shape of the metal is changed by hammering. Lead is a malleable material, as it can easily be shaped by hammering, but is not ductile since it is not strong enough to withstand a load if attempted to be drawn into wire. Heat may be used to make a material more malleable.

Elasticity

A material which is elastic will return to its original dimensions after being subjected to a load. If loaded above a point known as the elastic limit, the material will not return to its original dimensions and will be permanently deformed when the load is removed. Elasticity is essential in materials used in the manufacture of springs.

Toughness

A material is tough if it is capable of absorbing a great deal of energy before it fractures. A tough material will withstand repeated flexing or bending before it begins to crack or break. The working parts of power presses must be tough to withstand the repeated blows in pressing operations.

13.3 Comparison of properties

The mechanical and physical properties of common plastics and metallic materials are compared in Table 13.1 (page 194).

13.4 Plain-carbon steel

Plain-carbon steels are essentially alloys of iron and carbon together with varying amounts of other elements such as manganese, sulphur, silicon, and phosphorus. These additional elements are found in the raw materials used in the steel-making process and are present as impurities. Both sulphur and phosphorus are extremely harmful and cause brittleness in the steel – they are therefore kept to a minimum. The effect of these is offset by the presence of manganese. The carbon content varies up to about 1.4%, and it is this carbon which makes the steel harder and tougher and able to respond to the various heat-treatment processes.

Low-carbon steels cover a range of steels with carbon content up to 0.3%. These cannot be hardened by direct heating and quenching, but can be case-hardened. Steels containing 0.2% to 0.25% carbon, referred to as mild steels, are used in lightly stressed applications and can be readily machined and welded. They are used for general engineering purposes as bar, plate, sheet, and strip and for cold-forming operations. The tensile strength of rolled section for 0.2% carbon content is 300 N/mm^2, and for 0.25% carbon content increases to 430 N/mm^2.

Medium-carbon steels cover a range of steels with a carbon content above 0.3% up to 0.6%. They can be hardened by direct heating and quenching, and tempered to improve the mechanical properties. Steels containing 0.4% carbon are used where higher stressing and toughness is required for forgings, levers, shafts, and axles. In the normalised condition the tensile strength is around 540 N/mm^2, and in the hardened and tempered condition it can increase to around 700 N/mm^2.

Those steels with 0.6% carbon have a higher tensile strength, 700 N/mm^2 in the normalised condition, and can be hardened and tempered up to 850 N/mm^2. They are

Table 13.1 Comparison of mechanical and physical properties of common plastics and metallic materials

	Density (kg/m³)	Tensile strength (N/mm²)	Coefficient of linear expansion (10⁻⁶°C)	Specific heat capacity (J/(kg°C))	Thermal conductivity (W/(m°C))	Melting point (°C)	Resistivity (Ωm)
Mild steel	7800	505	15	463	47	1495	16×10^{-8}
Grey iron	7000–7300	150–400	11	265–460	44–52	1100	10×10^{-8}
Malleable iron	7300–7400	280–690	11	520	40–49	1100	–
S.G. iron	7100–7200	370–800	11	460	32–36	1100	–
Copper	8900	216	17	386	385	1083	1.7×10^{-8}
70/30 brass	8530	320	20	379	117	935	6.2×10^{-8}
Phosphor bronze	8820	400	18	379	70	1000	9.5×10^{-8}
Aluminium	2700	80	24	965	240	660	2.6×10^{-8}
Aluminium alloy	2790	250	22	965	150	600	4×10^{-8}
Zinc alloy	6700	280	27	418	113	400	5.9×10^{-8}
Lead	11 300	15	29	126	35	327	21×10^{-8}
Platinum	21 450	350	9	136	69	1773	11×10^{-8}
Silver	10 500	15	19	235	419	960	1.6×10^{-8}
Gold	19 300	120	14	132	296	1063	2.4×10^{-8}
						Softening point (°C)	
Polyethylene:							
l.d.	925	7–16	160–180	2300	0.34	85–87	10^{14}
h.d.	950	21–38	110–130	2220	0.46–0.52	120–130	10^{14}
P.V.C.	1390	58	50	840–2100	0.14	82	10^{14}
Polystyrene	1055	34–84	60–80	1340	0.11–0.14	82–103	10^{11}
ABS	1100	17–62	60–130	1380–1680	0.062–0.36	85	1.2×10^{13}
Acrylic	1200	48–76	50–90	1470	0.17–0.25	80–98	10^{12}
Polypropylene	900	29–38	110	1930	0.14	150	10^{14}
Nylon 6.6	1140	48–84	100–150	1680	0.22–0.24	75	$0.45 – 4 \times 10^{12}$

used where wear properties are of greater importance than toughness, for sprockets, machine-tool parts, and springs.

High-carbon steels cover the range of steels with a carbon content between 0.6% and 1.4%. Those containing 0.6% to 0.9% carbon are widely used in the hardened and tempered condition for laminated and wire springs and in the manufacture of spring collets. Steels having a carbon content above 0.9% are used in the hardened and tempered condition for hand and cutting tools where hardness is important, e.g. cold chisels, punches, files, and woodworking tools.

13.5 Heat treatment of plain-carbon steel

Heat treatment is a process in which a metal, in its solid state, is subjected to one or more cycles of heating and cooling in order to obtain certain desired mechanical properties. The mechanical properties listed in Section 13.2 can be altered by changing the size, shape, and structure of the grains from which the material is made up.

When a plain-carbon steel is heated through a sufficient temperature range, there is a particular temperature at which the internal grain structure begins to change. This temperature, known as the lower critical temperature, is about 700°C and is the same for all plain-carbon steels, Fig. 13.3.

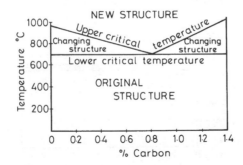

Figure 13.3 Relationship of critical temperature and carbon content

When the steel is heated still further, the structural changes continue until a second temperature is reached where the change in the internal structure of the steel is complete. This temperature is known as the upper critical temperature and varies for plain-carbon steel according to the percentage carbon content, Fig. 13.3.

The influence of the carbon content is so great that the heat treatment and subsequent employment of the steel is determined by this factor.

The temperature range between the lower and upper critical temperatures is known as the critical range.

If the steel, at a point above its upper critical temperature, is plunged into a cold liquid – a process known as quenching – the result will be to permanently fix this new structure, i.e. the structure is suddenly 'frozen' before it can change back to its original state.

If the steel is heated to above its upper critical temperature and instead of being quenched is allowed to cool slowly, structural changes take place in the reverse order to those during heating. When cold, the steel will have returned to its normal structure.

It follows from this that, if the rate of cooling is varied, considerable changes in structure and therefore variations in mechanical properties can be obtained.

It should be noted that the steel is still in a solid state during the heating and cooling cycles. The changes which take place are internal structural changes only – the steel never reaches its melting point.

Heat-treatment operations in industry are carried out in correctly controlled furnaces, the most common processes being annealing, normalising, hardening, and tempering.

Annealing

This process is carried out to soften the steel so that it may be machined or so that additional cold-working operations such as pressing and bending can be carried out.

The process involves heating the steel to a temperature depending upon its carbon content, Fig. 13.4, holding it at this temperature for a period of time depending upon the thickness of the steel, so that the whole mass reaches the correct temperature (known as 'soaking'), and finally allowing the steel to cool as slowly as possible. This slow rate of cooling is achieved by switching off the furnace, allowing the steel in the furnace and the furnace itself to cool at the same slow rate.

The result is a steel having large grains in its structure which is soft, ductile, low in strength, and easily shaped by machining, pressing, and bending.

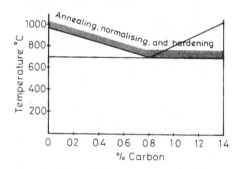

Figure 13.4 Relationship of heat-treatment temperature and carbon content

Normalising

This heat-treatment process is carried out to give the steel its 'normal' structure. For example, a steel which has been forged has a grain structure which has been distorted due to the hot working. Such a steel requires normalising, to return the grains to their normal undistorted structure to be in the best condition for use.

The process differs from annealing only in the rate of cooling. The steel is heated to the required temperature, depending again upon the carbon content, Fig. 13.4, and is allowed to soak. The steel is then removed from the furnace and is allowed to cool in still air. This gives a faster rate of cooling than annealing, resulting in a steel with smaller grains which is stronger but less ductile than an annealed steel.

Hardening

In contrast to annealing, this heat-treatment process is designed to produce a steel that is hard. Steel will vary in hardness depending upon the carbon content, the

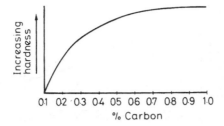

Figure 13.5 Hardness obtainable with a quenched plain-carbon steel

hardness increasing as the carbon content increases, Fig. 13.5. As the hardness increases so will the brittleness, which must be borne in mind when deciding what the material is to be used for.

Hardening is carried out by raising the temperature, again depending on the carbon content, Fig. 13.4, in the same way as for annealing and normalising, and allowing the steel to soak. The difference is again in the rate of cooling. This time the steel is removed from the furnace and is cooled very quickly, or quenched, by immersion in a suitable liquid such as water or oil.

Plain-carbon steels having a carbon content below 0.3% cannot be effectively hardened this way, due to the small amount of carbon present. These low-carbon steels can be hardened by a method known as case-hardening. The steel is heated to above the upper critical temperature while in contact with a carbon-rich material. Carbon is absorbed into the surface of the steel, raising the carbon content at the surface to around 0.9%, a process known as *carburising*. The steel can then be hardened as a 0.9% steel as previously described.

Carburising can be carried out using solid or liquid materials or a gas. Pack carburising is carried out by packing the steel in a solid carbon-rich material similar to charcoal in a steel box with the lid sealed using fireclay. The box is put in a furnace and the temperature is raised to between 900°C and 950°C and kept at this level for a period depending upon the required depth of penetration of the carbon. At the end of this period, the furnace is switched off and the box and its contents are allowed to cool slowly. Thus the steel now consists of a core with the original low-carbon content, usually 0.15%, and an outer case of high carbon content to the required thickness.

Subsequent heat treatment is carried out which produces a tough core (0.15% carbon) and a hard wear-resisting outer case (0.9% carbon). This method is used for shafts and spindles which, as well as requiring a hard wear-resistant outer surface, also require toughness to resist bending and torsional loads.

Tempering

This heat-treatment process is carried out after a steel has been hardened. Steel in a hard state is brittle and if used could easily break. Tempering removes some of the hardness, making the steel less brittle and more tough. This is done by reheating the hardened steel to a temperature between 200°C and 450°C, to bring about the desired structural change. The steel can then be quenched or cooled slowly, since it is the temperature to which the steel is raised which brings about the necessary structural change, not the rate of cooling. In general, the higher the temperature to which the hardened steel is raised the tougher the steel will be, with a corresponding reduction in hardness and brittleness, Fig. 13.6.

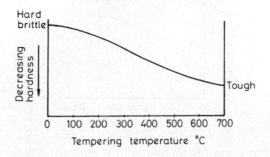

Figure 13.6 Relationship of tempering temperature and hardness

A component such as a scriber which requires to be hard and will not be subjected to shock loads would be tempered by reheating to a lower temperature. The head of a hammer which requires to be tough to withstand shock loads, with no great degree of hardness, would be tempered by reheating to a higher temperature.

Industrial tempering is carried out in correctly controlled furnaces, but it is a curious phenomenon that the colours caused by surface oxidation as steel is heated correspond fairly closely to the temperature within the tempering range, shown in Table 13.2.

Table 13.2 Tempering temperatures and oxide colours

Temperature (°C)	230	240	250	260	270	280	300
Oxide-film colour	Pale straw	Dark straw	Brown	Brownish purple	Purple	Dark purple	Blue

If, after hardening, a steel is rubbed with emery cloth to give a bright surface and is then reheated, the bright surface will take on the oxide colours – pale straw at the lower end, approximately 230°C, through to blue as the temperature is increased to approximately 300°C. These colours cannot of course be relied upon to give accurate results, but they may be a useful aid in the workshop.

More accurate results can be obtained by the use of thermal crayons. A crayon of an appropriate colour is selected for the desired temperature and is rubbed on the surface of the cold metal. The metal is then heated and the crayon deposit changes to a different colour, in accordance with a colour chart provided, when the correct temperature is reached.

13.6 Cast iron

There are many types of cast iron available, covering a wide range of mechanical and physical properties. All of them can be easily cast into a variety of simple or complex shapes. Like steel, cast iron is an alloy of iron and carbon, but with the carbon content increased to between 2% and 4%.

Grey cast iron

Grey cast iron is widely used for general engineering applications, due to its cheapness and ease of casting. It is available in a variety of different grades, from

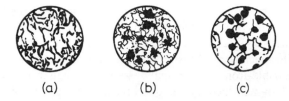

Figure 13.7 Cast irons; (a) grey, (b) malleable, (c) spheroidal-graphite

those that are relatively soft and of low strength to relatively high-strength harder materials. Tensile strengths range from 150 to 400 N/mm^2.

Grey cast irons contain carbon in the form of graphite flakes distributed throughout, which create a weak structure, Fig. 13.7(a). Other elements present include:

- silicon, which aids in the formation of the free graphite and may be present up to about 3%;
- phosphorus, which helps make the cast iron more fluid and may be present up to 1.5%;
- sulphur, present as an impurity from the furnace – too much sulphur tends to produce unsound castings so the sulphur content is kept to a minimum, about 0.1%;
- manganese, which toughens and strengthens the iron partly, because it combines with the unwanted sulphur, and is present up to about 1.5%.

Grey irons are brittle materials, have a high compressive strength between three and four times the tensile strength, and can be easily cast to shape and easily machined.

There is no limit to the size of grey-iron castings. Although not a corrosion-resistant material, grey iron does have a useful resistance to chemicals, water, gas, and steam, and for this reason is used for valves, pipes, and fittings with these substances.

Grey iron has a good wear-resistance and is used for castings in machine tools containing slideways where continuous lubrication is not always possible. The free graphite on the surfaces acts as a temporary lubricant and, when used up, creates minute pockets which then act as small reservoirs for the machine lubricant.

The reasonable mechanical strength and good thermal conductivity make grey iron a suitable material for cylinder heads, and brake and clutch applications in motor vehicles.

Malleable cast iron

All grades of malleable iron start as white-iron castings, free from graphite. The castings are then made tough and machineable by an annealing process. Due to the limited capacity of heat-treatment equipment, malleable iron castings are usually light in section and seldom weigh more than 50 kg.

Malleable irons are used in place of grey irons because of their increased tensile, impact, and fatigue strengths.

The three main types of malleable iron are: *whiteheart, blackheart*, and *pearlitic*. These have different properties due to the different heat treatments given to the initial white iron resulting in rosettes or clusters of graphite, Fig. 13.7(b).

Whiteheart-malleable-iron castings are used where there is a need for small ductile components, particularly with thin sections. Uses include pipe fittings, frame sockets, steering-column housings and engine bearers on motor cycles, and parts for agricultural and textile machinery. Tensile strengths range from 340 to 410 N/mm^2.

Blackheart-malleable-iron castings have somewhat better tensile strengths than grey iron but much higher ductility and good machineability. They can be used under shock-loading conditions and are widely used in the car industry for small components such as door hinges and brackets where wear-resistance is not a primary requirement. Tensile strengths range from 290 to 340 N/mm^2.

Pearlitic-malleable-iron castings are used where high strength is required together with a good wear-resistance. There are many applications, particularly in the car industry, which include axle and differential housings and gears. Tensile strengths range from 440 to 690 N/mm^2.

Spheroidal-graphite cast iron

Spheroidal-graphite cast iron combines the strength, toughness and ductility of steel with the ease of casting of grey cast iron. Usually abbreviated to SG iron, it is also known as 'ductile' or 'nodular' iron.

In SG iron, the graphite is present as spheroids or nodules, Fig. 13.7(c), which are induced by adding magnesium before casting. This gives values of mechanical properties higher than with most other cast irons.

There is no practical limit to the section size or mass of SG iron castings, which leads to their use in a wider variety of applications than the malleable irons.

In many instances SG iron is used as a substitute for steel and replaces forgings, the advantages of the cast shape and reduced machining making it economical. Uses include pipes and fittings for gas, oil, water, sewage, and chemicals. In motor vehicles the uses of SG iron include cylinder blocks, crankshafts, connecting rods, exhaust manifolds, water pumps, timing gears, gearbox casings, steering boxes, and many other items. In agriculture the uses include a variety of tractor components, e.g. front-axle, steering, and suspension units, transmission housings, and drive gears, as well as equipment such as plough shares, cultivator discs, and mowing-machine parts.

SG iron is widely used in civil, mining, and power engineering, construction, the steel industry, and in the manufacture of machine tools.

The tensile strength varies over a wide range from 370 to 800 N/mm^2.

13.7 Copper and its alloys

Copper

Copper is a soft ductile material which increases in hardness and strength when cold-worked, i.e. in bending, spinning, and drawing. The main advantages of copper are its high thermal and electrical conductivities and excellent corrosion-resistance to chemicals, water, and the atmosphere.

The electrical and thermal conductivities of high-purity copper are greater than those of any other metal except silver. Many grades of commercial purity are available.

Tough-pitch copper: Tough-pitch copper, of 99.85% purity, is used in chemical and general engineering applications where the highest conductivity is not required. Where the highest conductivity is required for conductors and electrical components, a highly refined grade known as tough-pitch high-conductivity copper, of 99.9% purity, is used.

Phosphorus deoxidised arsenical copper: Phosphorus deoxidised arsenical copper, of 99.2% purity with 0.3% to 0.5% arsenic, is widely used for copper tube and general engineering applications where brazing and welding are required.

Brass: Brass is essentially an alloy of copper and zinc, but may also contain small amounts of other alloying elements to improve strength, corrosion-resistance, and machining characteristics.

70/30 brass: This alloy contains 70% copper and 30% zinc and is highly ductile. It is often referred to as cartridge brass, due to its use in the manufacture of ammunition. Because of its ductility it can be used in pressing, spinning, and drawing operations.

The addition of 1% tin at the expense of zinc gives better corrosion-resistance, and the resulting material, known as Admiralty brass, is used for condenser tubes.

60/40 brass: This alloy contains 60% copper and 40% zinc and is used where the material is to be hot-worked. It is ideal for producing hot stampings and extruded bars, rods, and sections.

To produce a brass with excellent high-speed machining properties, known as free-cutting brass, 2% to 3% lead is added.

High-tensile brass: This alloy is basically a 60/40 brass with additional alloying elements such as tin, iron, manganese, and aluminium. The effect of these alloying elements is to increase the strength and corrosion-resistance.

This material can be cast, forged, and extruded. Uses include high-pressure valves, components for pumps, and marine propellors; non-sparking tools for the gas, oil, and explosives industries; and in the manufacture of nuts, bolts, and studs.

Bronze

Bronze is essentially an alloy of copper and tin, but may also contain additional elements such as zinc and phosphorus. Bronzes containing copper, tin, and phosphorus are known as phosphor bronze, while those containing copper, tin, and zinc are known as gunmetal.

Phosphor bronze: Phosphor bronze is available in a wrought condition as sheet, strip, plate, rod, bar, wire, and tube and also as cast materials.

As a wrought alloy in the form of wire and strip containing 5% tin and up to 0.4% phosphorus with the remainder copper, the alloy can be heat-treated. In the heat-treated condition it has good elastic properties as well as corrosion-resistance and is used for springs.

As a cast material containing a minimum of 10% tin and 0.5% phosphorus with the remainder copper, it is used for bearing applications.

Gunmetal: The most common material of this type is Admiralty gunmetal, containing 88% copper, 10% tin, and 2% zinc. This alloy casts extremely well and has a high resistance to corrosion with good mechanical properties.

Gunmetal castings are largely used for naval purposes – for valves, pump bodies, and fittings used with water and steam.

202 Workshop Processes, Practices and Materials

Up to 5% lead may be added to improve pressure tightness, ease of casting and machining, and antifriction properties.

13.8 Aluminium and its alloys

Advantages

Pure aluminium is light, soft, ductile, corrosion-resistant, and highly conductive to heat and electricity. It is used in its pure form where strength is not of major importance, e.g. as foil for packaging. Alloying with other elements such as copper, magnesium, manganese, silicon, and zinc increases the strength and hardness and enables the material to be heat-treated to give additional properties.

One of the most important characteristics of aluminium and its alloys is the thin oxide film which forms on their surfaces when exposed to the atmosphere. If the oxide film is broken it will reform quickly, and this gives these materials excellent corrosion-resistance. This oxide film can be artificially thickened to give added protection – a process known as anodising – and can be easily coloured to provide a highly decorative appearance.

Forms of supply

Aluminium and its alloys are available in a wide range of shapes and forms.

- *Foil* – in thicknesses from 0.2 mm to 0.005 mm, usually from aluminium of 99% purity.
- *Sheet* – available in standard sizes 1000 mm × 2000 mm and 1250 mm × 2500 mm in thicknesses from 0.5 mm to 3.0 mm.
- *Strip* – available in widths of 500 mm, 1000 mm, and 1250 mm in thicknesses from 0.25 mm to 2.0 mm.

Both sheet and strip are available with the surface prepainted in a range of colours, used in the manufacturer of caravans.

- *Plate* – defined as having a minimum thickness of 3 mm, available in a range of sizes.
- *Bars* – defined as round, square, rectangular, and polygon solid section in sizes greater than 6 mm and supplied in straight lengths.
- *Extruded sections* It is possible to produce an enormous variety of shapes by extrusion. The cost of the extrusion die is high, and, to be economical, large quantities of section have to be produced. Although some standard extruded sections are available, such as angles, the majority are made to a customer's own requirements.
- *Tube* – may be extruded, drawn, or seam-welded from strip in a variety of sizes.
- *Wire* – available in sizes up to 10 mm diameter, wire is used in the production of rivets, nails, screws, bolts, and welding rods, in metal spraying, and for electrical cables.
- *Forgings* A number of alloys forge very well – a process originally developed for aircraft components where high-strength and low-weight properties were required.
- *Castings* Where intricate shapes are required, the range of cast alloys can be used. These may be cast by sand, gravity-die-, and pressure-die-casting methods and with other more specialised processes.

Applications

The uses of aluminium and its alloys are virtually limitless, covering the fields of transport; electrical, structural, civil, and general engineering; household items; packaging in the chemical and food industries; and many more.

Applications in transport include cladding and floor sections of commercial vehicles, prepainted sheet for caravans, the superstructure of ships and hovercraft, and a variety of components in aircraft.

Electrical engineering uses include large and small cables, foil, and strip windings for coils and transformers.

Structural and civil engineering uses include roofing and structural applications, door and window frames, and a variety of decorative items.

General engineering uses include watches, photographic equipment, machinery for textiles, printing, and components for machine tools.

Household items using aluminium include electrical appliances, pans, and furniture; and packaging uses foil for food, drink, tobacco, and chemicals.

13.9 Die-casting alloys

The process of die-casting uses a split metal mould into which molten metal is:

- poured under the action of gravity (gravity die-casting); or
- forced in under pressure (pressure die-casting).

When the metal solidifies, the split die is opened and the casting is ejected.

The common metals used in die-casting are zinc- and aluminium-based alloys.

Zinc alloy containing 4% aluminium with the remainder zinc is most widely used for the production of castings required in large quantities, accurately, and with a good surface appearance. This alloy is cast at temperatures around 400°C and is very fluid in the molten state, which leads to its use for intricate shapes and thin sections.

Zinc alloys can be easily electroplated or painted to give a decorative or protective finish. Uses include the mass production of car parts such as door handles, lock parts, carburettors, fuel pumps, and lamp units, as well as camera parts, power tools, clock parts, and domestic appliances.

Aluminium alloys, typically containing 3.5% copper and 8.5% silicon, with the remainder aluminium, are less tough than zinc alloy but have the advantage of low density. Aluminium alloys require casting temperatures of around 650°C. Their many uses include car crankcases, gearboxes, timing-case covers, and components for electrical and office equipment.

13.10 Lead

Lead in its pure state is very soft and has low mechanical strength. In this form it is used widely in the chemical industry, due to its high corrosion-resistance. Because of its low mechanical strength it is applied in the form of a lining to other, stronger materials. It is also used for radiation shielding. When alloyed with antimony, the strength and hardness is increased and it is used for the production of lead bricks for nuclear shielding. Two of its largest uses are as cable sheaths for power cables and in the manufacture of connectors and grids in lead–acid batteries.

Lead–tin alloys give a range of soft solders.

13.11 Contact metals

Contacts are used where there is a need to make and break electrical connections. They must be reliable when working under a variety of conditions such as heat, cold, humidity, vibration, dust, and corrosive atmospheres, and must give a long service life against mechanical wear, heat, fatigue, metal transfer, and corrosion.

Platinum

Pure platinum has two important properties as a contact material: it has a high melting point, 1770°C, and is therefore able to resist the effects of the electrical current arcing across the faces; also it has a high resistance to corrosion.

Pure platinum is relatively soft and is used in light-duty applications on sensitive relays and instruments. When alloyed with iridium, greater hardness and mechanical properties can be obtained, leading to its use in medium-duty applications.

Silver

Silver has the highest thermal and electrical conductivities of any metal and is resistant to corrosion. Pure silver is used on contacts in the form of an electroplated film. It is used in a variety of light- and medium-duty applications in telephone relays, sliding contacts, thermostats, and voltage regulators. When alloyed with copper, the electrical conductivity is lowered and the hardness and mechanical properties are increased.

Gold

The high electrical conductivity of gold, surpassed only by silver and copper, together with its resistance to tarnishing, makes it suitable as a contact material. Alloying with copper and silver gives good resistance to wear and is used for sliding contacts in light-duty applications.

A 5% nickel/gold alloy is widely used.

Tungsten

Tungsten has the highest melting point of any known metal: 3380°C. It is alloyed with other metals in powder form by a process known as sintering. Alloying with silver or copper gives varying degrees of conductivity, hardness, and wear-resistance.

13.12 Bearing materials

Plain bearings are the oldest form of bearing. They may be manufactured completely from a single bearing material or composition of materials, or, alternatively, a thin layer of bearing material can be attached to a backing which is usually of a stronger material.

Where a metal bearing material is used with a metal shaft, metal-to-metal contact must be avoided to prevent seizure of the two metals. This is done by providing a film of lubricant between the two surfaces. Where it is not possible to provide a film of lubricant, a non-metallic bearing material will have to be used.

Ideally a bearing material should possess the following properties:

- good thermal conductivity, to carry heat away from the bearing;

- sufficient strength to carry the loading of the shaft or sliding part without permanent deformation;
- resistance to corrosion by lubricants or the atmosphere;
- the ability to operate over a range of temperatures (melting point and coefficient of expansion are important factors);
- the ability to deform slightly, to compensate for small misalignments or surface irregularities;
- the ability to allow dirt, grit, or filings to embed in the surface rather than pick up and seize on the shaft or sliding part;
- the ability to resist wear.

Phosphor bronze

Cast phosphor-bronze bearings possess high strength to support heavy loads at low speeds and have excellent corrosion-resistance. The bearing and shaft must be accurately aligned.

Lubrication is required and must be plentiful and reliable – oil grooves are usually provided in the bore of the bearing to give good distribution over the complete surface.

Porous bronze bearings are available, made from powdered metal. The powder is pressed into shape leaving air spaces or pores. When the bearing is subsequently immersed in oil, these pores absorb oil. In operation, the oil seeps from the pores to provide an oil film which prevents metal-to-metal contact. Alternatively, graphite can be incorporated in the initial powder to provide a self-lubricating property.

Porous bronze bearings can be used at high speeds with light loading and offer good corrosion-resistance. They are designed to last the life of the assembly without additional lubrication and are used in domestic appliances, starter motors, and car parts.

Cast iron

Cast iron is not used as a bearing material for rotating parts: it has high load characteristics and is used mainly for sliding surfaces such as machine-tool parts, e.g. milling-machine slides, lathe beds, etc.

Because of the free graphite it contains, cast iron has certain self-lubricating properties but requires additional lubrication as there is a tendency for the material to pick up and quickly seize.

PTFE

This is a thermoplastic material, the correct name for which is polytetrafluoro-ethylene, though it is usually referred to as PTFE.

It possesses a wide temperature range of use from, $-200°C$ to $+250°C$, is resistant to most chemicals, and can be easily machined. Its main advantage as a bearing material is its low coefficient of friction.

However, PTFE has big disadvantages as a bearing material – it has marginal wear-resistance, poor thermal conductivity, and a high coefficient of expansion. These disadvantages can be reduced by the addition of fillers such as glass fibre, graphite, bronze, and molybdenum disulphide; alternatively, a film of PTFE on a steel backing provides strength, wear-resistance, good heat conductivity, and low thermal expansion.

It is used for bearings required to operate without lubricant and as static bearing pads for bridge expansion joints.

Nylon

Nylon is a thermoplastic material which can operate under light to medium loads, is tough, abrasion-resistant, resists chemical attack, is non-toxic, and can be used continuously at temperatures up to about 90°C.

Like other thermoplastic materials, nylon has a high coefficient of expansion and low thermal conductivity. It also readily absorbs moisture. The addition of fillers such as glass fibre, graphite, and molybdenum disulphide reduces these disadvantages and increases the wear-resistance and operating temperature.

Uses of nylon bearings include bushes in domestic appliances; bushes, thread guides, rollers, and slides in textile machinery; ball joints; and bearings in suspension and steering systems in cars.

Graphite

Graphite is used as an additive for bearings required to operate without lubricant. It is added to PTFE and nylon and gives a lower coefficient of friction than is possessed by the materials themselves.

The antifriction properties of graphite can be illustrated by sliding a block of metal across a surface plate. Now sprinkle some shavings from a pencil lead containing graphite on the surface plate and again slide the block across the surface. It will now slide very much more easily.

13.13 Metal protection

In 1971 a Government committee estimated that corrosion was costing Great Britain nearly 1400 million pounds a year and that at least 300 million pounds of this could be saved by applying known corrosion-control techniques. These figures include not only direct loss of material but also other indirect factors, such as loss of production, reduction in plant efficiency, and contamination of products resulting from corrosion.

Corrosion will occur and can be controlled. It is therefore essential that engineering technicians are aware of the causes of corrosion and of the methods available for its control.

One of the most common methods used in industry for the control of corrosion is to apply a protective coating to material surfaces. The type of protective coating and the method of application depend on a wide variety of factors which include operating conditions, the immediate environment, whether indoors or outdoors, the likely risk of damage, and appearance.

13.14 Corrosion

Corrosion is an electrochemical process in which a metal reacts with its surroundings to form an oxide or compound similar to that of the ore from which it was extracted.

Metals vary greatly in their corrosion-resistance – chromium and titanium have good resistance, while steel readily corrodes. The oxide film formed on chromium and titanium closely adheres to the surface and protects the metal from further oxidation. In the case of steel, the oxide film in the form of rust is loose, allows moisture to be retained, and promotes further corrosion. If corrosion is allowed to continue, the steel will eventually be completely consumed, i.e. the metal will have returned to the condition of the ore from which it was extracted.

The electrochemical process is caused by a series of cells known as corrosion cells. A cell is made up of:

- an *anode* – an area where corrosion takes place;
- an *electrolyte* – an electrically conducting solution;
- a *cathode* – which completes the cell and is not consumed in the corrosion process.

The anode and cathode may be two surfaces in contact with each other or they may occur on the same surface due to small variations in the metal structure, Fig. 13.8.

Figure 13.8 Corrosion cells

The electrolyte is usually water – present as moisture, rain, or sea water – which may also contain elements of dust and gases which accelerate the corrosion process. The metal may be in constant contact with the electrolyte, e.g. underground structures and liquids in pipes, tanks, and various vessels; alternatively, the metal may be indoors subjected to differing degrees of humidity or dampness or outdoors in all weather conditions. The rate of corrosion is influenced by the electrical conductivity of the electrolyte; i.e. high rate in salt solutions, low rate in high-purity water.

Corrosion can be controlled by using a material having good resistance. Unfortunately, materials which have good resistance to corrosion are more expensive than those with poor resistance. It is seldom possible or even desirable to replace all materials with high-cost corrosion-resistant ones, and other methods have to be considered.

As already stated, corrosion can take place only if an electrolyte is present. The obvious method of controlling corrosion, then, is to prevent electrolyte from contacting the metal surfaces; i.e. exclude the environment from the metal. One method of doing this is to provide a protective coating on the metal surfaces, the method chosen depending on the type of metal to be protected, the environment in which the metal will operate, and the coating material.

13.15 Protective coatings

Electroplating

Electroplated coatings are commonly applied to meet a variety of service requirements, and also for decorative purposes. Service requirements include corrosion-resistance, wear-resistance, and contact with chemicals and foodstuffs. The coatings described here are those used chiefly for their corrosion-resistant properties.

Cadmium plating: Cadmium has excellent corrosion-resistance, especially in marine applications as a coating on iron and steel. Its main use is with military and sea-going equipment and in steel components on aircraft. For industrial use it is now being replaced by zinc coatings, which are cheaper.

Cadmium is poisonous and must not be used on articles or equipment for use in contact with food or drinking water. It should not be applied to articles which may be subjected to temperatures in excess of 250°C, as poisonous fumes are then given off.

Chromium plating: Bright chromium plate is the most widely used electroplated coating. It combines resistance to corrosion, wear, and heat, use in contact with foodstuffs, and decorative qualities.

To obtain bright chromium plate, the chromium is deposited over a deposit of bright nickel. The deposit of nickel, which itself has good corrosion-resistance, gives protection to the base metal and is itself protected from surface oxidation by the chromium. Black chromium plating is used on windscreen wipers, bumpers, and automobile parts to provide corrosion-resistance with a non-glare safety finish.

Nickel plating: Nickel deposits are most widely used as a base for chromium in producing bright chromium plate. Apart from its decorative appeal, nickel is used as a hard wear- and corrosion-resistant coating on steel and on copper, zinc, and aluminium alloys.

Tin plating: Tin plating is employed as a protective coating on both ferrous and non-ferrous metals, particularly where the product will come in contact with foodstuffs. It is also used for electrical components, for protection against corrosion and as an aid to soldering.

Zinc coating: The best-known zinc coating is galvanising, which is a hot-dip process almost exclusively used on industrial, constructional, and domestic goods which require a thick coating.

Zinc plating is used on many steel components which were formerly cadmium-plated. Zinc is unsuitable for electroplating articles likely to come in contact with foodstuffs, since it dissolves in dilute acids and alkalis.

Anodising

Anodising is used extensively for the production of protective and decorative films on aluminium and its alloys. In the anodising process, the naturally occurring film of aluminium oxide is artificially thickened by an electrolytic process to give increased corrosion-resistance. This is done by making the article the anode immersed in a weak solution of sulphuric or chromic acid.

The thickness of film depends on the electrical current and the process used and varies from 2.5 μm to 25 μm. Anodising provides a decorative finish where the surface may be dyed to give a wide range of attractive colours.

Chemical coatings

Metal surfaces can be treated chemically with an appropriate solution which provides a coating with limited protection against corrosion. These coatings also provide an excellent base for paint. The combined effect of the chemical and paint coatings gives an extremely high degree of corrosion-resistance.

Phosphating: This process is applied mainly to steel components, which are sprayed or dipped in a phosphoric-acid solution containing iron, zinc, or manganese phosphate. The most common process uses zinc phosphate and is employed in the treatment of car bodies and domestic appliances before painting. Where long-term corrosion-protection is required, the phosphate coating must be sealed with oils, lacquer, or paint.

Chromating: This process is applied to magnesium alloys used in aircraft, the armed services, and car wheels. The coating is produced by dipping the component in a solution of sodium or potassium dichromate with other additions depending on the method used. The coating provides improved corrosion-resistance and adhesion for a paint coating.

Paints

A wide range of paints is available for industrial use, and their primary function is to give added corrosion-resistance to the metal surface as well as providing a decorative appearance.

Paint consists of a mixture of pigment, which gives body and colour, and a liquid known as the binder which becomes solid during the drying process. To reduce the viscosity so that the paint can be easily applied, solvents or thinners are added. These solvents and thinners evaporate during drying and do not form part of the paint film.

To provide the maximum protection, a paint system should ideally consist of three types of paint:

- *primer* – applied to the clean surface and designed to adhere strongly and prevent corrosion;
- *undercoat* – applied as an intermediate film to provide a basis for the correct final colour and either matt or low-gloss to provide a good 'key' for the finishing paint;
- *finishing paint* – the choice of this paint will depend on the degree of protection and the decorative effect required.

Alkyd-resin paints: Alkyd resins are produced by reacting glycerine with certain acids. Modified with vegetable oils, they provide the basis for a range of tough air-drying paints with excellent adhesion which dry relatively quickly to a high-gloss finish. These resins form the basis of most decorative and maintenance paints and are widely used in undercoats and primers.

Industrial uses are for agricultural equipment and machinery too large for stoving which require a quick, hard-drying, solvent-resistant finish. When mixed with urea and melamine resins, the alkyd resins combine toughness and durability with low cost and constitute the majority of industrial stoving enamels for refrigerators and domestic equipment.

Chlorinated-rubber paints: Chlorinated rubber is produced by the chemical action of chlorine on raw rubber. Modified with plasticisers and with added solvents, it produces an air-drying film having outstanding resistance to chemical attack and to penetration by water. For these reasons, chlorinated-rubber paints are used in breweries, chemical plants, bridges, cranes, and ships.

Acrylic paints: Paints based on acrylic resins are widely used, particularly for motor-vehicle applications. Additional heat-curing resins can be incorporated which require stoving to bring about the chemical reaction and speed the drying process. Their advantages are speed of drying to give tough films with excellent durability and colour retention.

When used on car bodies, these paints require a stoving time of about one hour at 130°C.

Epoxy resins: Epoxy-resin-based paints are used where a high degree of chemical resistance is required. Stoving temperatures of about 200°C are required, with a drying time of 20 minutes.

Summary of metal protection

Table 13.3 summarises the various methods of metal protection.

Table 13.3 Summary of metal-protection treatments

Protective treatment	Material treated	Reason for treatment
Cadmium plating	Steel	Corrosion-resistance
Chromium plating	Steel and non-ferrous	Corrosion-resistance; wear-resistance; heat-resistance; decorative finish
Nickel plating	Steel and non-ferrous	Corrosion-resistance; wear-resistance; undercoat for bright chromium plating; decorative finish
Tin plating	Steel and non-ferrous	Corrosion-resistance; chemical resistance (foodstuffs); aid to soldering
Zinc coating	Steel	Corrosion-resistance; decorative finish
Anodising	Aluminium and aluminium alloys	Corrosion-resistance; wear-resistance; coloured decorative finish
Phosphating	Steel	Corrosion-resistance; undercoat for painting
Chromating	Aluminium and magnesium alloys	Corrosion-resistance; undercoat for painting
Paint	All	Corrosion-resistance; decorative finish

13.16 Painting

As with all other methods of protection, surfaces to be painted must be free from dirt, rust, and grease and wherever possible pretreated by phosphating or chromating to improve adhesion. In selecting the type of paint, consideration must be given to the method of application.

Brushing

Applying paint by brush has the disadvantage of being slow. Unskilled operators can be used where the finished surface is not critical, but where intricate and high-class work is essential a skilled operator is required. Brushes are expensive and need careful handling and proper cleaning after use.

Spraying

This method was developed for use with the quick-drying cellulose paints which dried too quickly for use with a brush.

The simplest form of spray gun uses compressed air. The paint, held in a container, is sucked up, atomised by the stream of compressed air, and blown through a nozzle.

Airless spraying is a method used especially where heavy paint coatings are required. A high-pressure pump is used to carry the paint to a very fine nozzle on the spray gun. The paint atomises after leaving the nozzle and produces a 'soft' spray, as there is no accompanying blast of compressed air through the nozzle. The resultant spray is much finer, applied with less force, produces less overspray, and is twice as thick as with compressed-air spraying. This method can be readily applied to automatic spraying systems.

Electrostatic spraying is carried out by electrically charging the atomised paint. As it leaves the gun, the paint is attracted to the work being coated, which is earthed. Paint which by other methods would pass round the edge or through holes is attracted to the surface, resulting in considerable paint saving. Tubular components can be painted all over by spraying from one side. This method can be used by hand or, more widely, in automatic installations in mass-producing industries in the manufacture of bicycles, car components, and domestic equipment.

Solvent-based paints give off toxic vapours and, especially with spraying, the atmosphere becomes laden with fine solid paint particles, presenting a fire and health hazard. Suitable systems have to be installed to continuously extract, filter, and cause air movement and replacement.

Dipping

This method is widely used in industry and consists of immersing the article in a tank of paint, slowly withdrawing it, and allowing it to dry either in air or by stoving. The quality of surface is not as good as with spraying, and there is a tendency for the paint film to be thicker towards the bottom of the article. To be suitable for dipping, an article must be open at the bottom, to allow excess paint to drip off.

This method can be completely automated and is used in the manufacture of motor cycles, motor-car chassis and bodies, and domestic radiators. The articles are suspended on a conveyor system, dipped, and withdrawn, before passing through an oven to provide quicker drying.

Paints containing highly flammable solvents cannot be used with this process in close proximity to an oven, due to the risk of fire. Water-based paints are used to overcome this hazard. However, where a hazard does exist, automatic fire-fighting devices are installed.

Roller coating

This method is used to apply paint continuously to large flat areas at high speed. The material, in the form of sheet or coil, is drawn between horizontal rollers, one of which is covered by a thin film of paint. Accurate uniform thickness of film can be achieved, which leads to economical use of paint.

Tumbling

This method is used where large numbers of small articles are to be painted. The articles are placed in a drum or barrel together with a quantity of paint and are rotated to give an even distribution on all surfaces. The articles are then removed for air drying or stoving.

Paint drying

Newly painted articles are in a wet condition which readily attracts dust. It is important that the paint film is dried as quickly as possible.

Air drying takes place at room temperature, and all possible steps should be taken to ventilate and prevent dust in the area. The introduction of warm air in an enclosure helps to speed up the drying process.

For high production rates, the drying operation must be speeded up by applying heat, usually at temperatures above 150°C in a drying oven. This drying operation, known as stoving, is carried out in a variety of types of oven, the heat transfer taking place by convection or radiation (infra-red). Infra-red-radiation stoving is more rapid than convection stoving using gas, oil, or electricity.

It is essential to allow an air-drying or 'flash-off' period of about ten minutes before stoving, so that some of the solvent may evaporate. Heat applied too quickly can result in bubbles or craters forming on the paint surface and can present fire hazards with highly flammable solvents.

14 Plastics

Most plastics used today are man-made and are described as synthetic materials, i.e. they are made by a process of building up from simple chemical substances.

Some plastics are soft and flexible, others hard and brittle, and many are strong and tough. Some have good thermal and electrical properties, while others are poor in these respects. Crystal-clear plastics are available, while others can be produced in an extremely wide range of colours. Most plastics can be easily shaped using heat or pressure or both.

14.1 Thermoplastics and thermosetting plastics

With a few exceptions, plastics are compounds of carbon with one or more of the five elements hydrogen, oxygen, nitrogen, chlorine, and fluorine. These compounds form a diverse groups of different materials, each with its own characteristic properties and uses. All plastics materials are based on large molecules which are made by joining together large numbers of smaller molecules. The small molecules, known as monomers, are derived from natural gas and crude oil and are subjected to suitable conditions to join up and form long-chain-molecular products known as polymers. The process of joining the molecules together is known as polymerisation, and the names of the plastics made in this way frequently contain the prefix 'poly'. For example, the monomer ethylene is polymerised to form very-long-chain molecules of the polymer polyethylene (also called 'polythene').

Plastics are mostly solid and stable at ordinary temperatures, and at some stage of their manufacture they are 'plastic', i.e. soft and capable of being shaped. The shaping process is done by the application of heat and pressure, and it is the behaviour of the material when heated that distinguishes between the two classes of plastics: thermoplastics and thermosetting plastics.

Plastics made up of molecules arranged in long chain-like structures which are separate from each other, soften when heated, and become solid again when cooled. By further heating and cooling, the material can be made to take a different shape, and this process can be repeated again and again. Plastics having this property are known as thermoplastics and examples include polyethylene (polythene), PVC, polystyrene, ABS, acrylics, polypropylene, nylon, and PTFE.

Other plastics, although they soften when heated the first time and can be shaped, become stiff and hard on further heating and cannot be softened again. During the heating process, a chemical reaction takes place which cross-links the long chain-like structures, thus joining them firmly and permanently together – a process known as curing. Plastics of this type are known as thermosetting plastics, or thermosets, and examples include phenolics ('Bakelite') and aminos (urea formaldehyde and melamine formaldehyde).

14.2 Types of plastics

Thermosetting plastics

Phenolics: Phenolic plastics are based on phenol-formaldehyde resins and are often referred to as 'Bakelite'. The material has a limited colour range and is supplied only in dark colours, principally brown and black.

Phenolics are good electrical insulators and show good resistance to water, acid, and most solvents. They are rigid and have low thermal conductivity, and for this reason are used as saucepan handles. The normal operating-temperature limit for phenolic mouldings is 150°C, but grades are available which will operate at up to 200°C for limited periods.

Phenolic materials contain fillers to modify properties such as heat-resistance and electrical insulation. Typical fillers are woodflour for general-purpose grades, asbestos for heat-resistance grades, and mica for good electrical properties. Typical uses include saucepan and kettle handles, terminal blocks, electrical switchgear, cooker control knobs, toilet seats, and voltage-regulator covers on cars.

Aminos: The aminos are products of urea or melamine with formaldehyde to form urea formaldehyde and melamine formaldehyde. The characteristics of these materials are:

- an unlimited range of colours;
- good electrical properties;
- resistance to oil, grease, and solvents;
- good heat-resistance;
- good strength and abrasion-resistance;
- hardness, rigidity, and durability;
- freedom from odour and taste.

In addition, melamine formaldehyde shows:

- improved resistance to hot and cold water;
- improved resistance to staining and discoloration from tea, coffee, or fruit juices.

Consequently melamine formaldehyde is used principally for moulded cups, saucers, plates, bowls, etc.

Urea formaldehyde is widely used in domestic electrical fittings such as switch covers, plug tops, socket bases, and lamp sockets, as well as for ornamental bottle tops for scent etc. and for toilet seats.

Glass-fibre-reinforced plastics (GRP): GRP is made up of a resin and layers of strong fibrous glass. The resin is usually polyester, and the glass-fibre reinforcing material is available in a variety of forms.

Glass fibre is made by rapidly drawing and cooling molten glass into the form of continuous filaments, which are bundled together to form strands immediately after drawing. These strands are then made into chopped-strand mat or into yarn for weaving into glass cloth or into a rope-like form known as rovings which in turn can be woven to produce a thick coarse cloth known as woven rovings.

Products are made in a simple mould which may be made of wood, plaster, or GRP itself. The mould surfaces must be accurately finished and well polished, to ensure a smooth surface on the moulding. Alternate layers of activated resin, plus any colouring pigment required, and glass reinforcement are laid until the required thickness is reached. The moulding is then left to cure before being removed from the mould.

The resins used should be worked with in a well ventilated area only, as concentrations of vapour from the solvents used can be harmful.

GRP is used to produce light durable tough constructions, any colour and any shape. It is used in the manufacture of boat hulls, canoes, lorry cabs, pipes, rocket missiles, car bodies, light fittings, roofing and building panels, doors, and parts for ships and aircraft, as well as many other products.

Thermoplastics

Polyethylene (or polythene): Depending on the process used in its production, polythene can be low density (LD) or high density (HD). The high-density type is stiffer and has a higher softening point.

Polythene is made in larger amounts than any other plastics. It is cheap, has a high resistance to water, oils, and chemicals, and can withstand considerable temperature changes.

Low-density polythene is widely used as film, bags, and sacks for packaging; to line containers too large to be made entirely of polythene; and for houseware, dustbins, and toys. The more rigid high-density polythene is used for large moulded containers, bottles, milk crates, tote boxes, pipe, sheet, and for better-quality houseware as it does not become limp if hot water is held in it.

Polyvinyl chloride (PVC): This is the second most widely used plastics and is one of the most versatile, being available in many forms with a wide range of hardness and flexibility. PVC possesses good physical strength, durability, and resistance to water and chemicals, and is a good electrical insulator. Plasticisers can be added to produce a flexible PVC, otherwise a rigid PVC is obtained.

Rigid PVC is used for pipes and fittings, roofing sheet, wall cladding, electrical conduit, ventilation ducting, gramophone records, and bottles for edible oil, sauces, and fruit drinks. Flexible PVC is used for cable and wire insulation, floor and wall coverings, coated fabrics for upholstery and clothing, and for hosepipes.

Polystyrene: This is the third most widely used plastics and is among the cheapest to produce. It is resistant to water, oils, and alkalis but is melted by solvents.

Polystyrene can be produced as very-thin-walled containers for dairy products, foodstuffs, and vending cups and is also used in packaging for eggs and meat. Polystyrene-based plastics are widely used in vacuum cleaners, fridge liners, and spools, cassettes, and cartridges for tape and photographic film. In the form of expanded foam, polystyrene is used for insulation as ceiling tiles and blocks, and as preforms in packaging.

Once ignited it will continue to burn, although self-extinguishing grades are available.

Acrylonitrile-butadiene-styrene (ABS): This material has exceptional impact strength, resistance to low temperatures and chemicals, and can be produced with an exceptionally high-gloss finish in a wide range of colours.

ABS is used in place of PVC and polystyrene where greater toughness is required. Its uses include instrument panels on cars, children's pedal cars, sailing and rowing boats, caravan parts, liners for refrigeration, and for gas pipes, where its toughness and low-temperature properties make it ideal for underground use. In addition, sheets can be produced with a grained finish for use in the manufacture of suitcases, machine guards, and furniture.

Polymethyl methacrylate (acrylic group of plastics, or Perspex): Perspex is best known in its self-coloured or clear form, as its main property is complete transparency with optical clarity. It also possesses exceptional stability to outdoor weather, resulting in wide use in signs, displays, and light fittings. At room temperature it is rigid, can be easily machined, and can be joined by cementing. At 160°C it becomes pliable and can be easily shaped.

Perspex can be highly polished and is used for reflectors, lenses, and for light transmission. It is also available in a vast range of colours and is widely used for sanitary ware, especially baths, furniture, record-player lids, and radio and speedometer scales.

Polypropylene: Polypropylene is one of the lightest plastics available. It is tough yet rigid, has a high resistance to chemical attack, will withstand repeated bending, and will operate at temperatures above 100°C.

Polypropylene is widely used in car interiors for fascia panels, glove boxes, kick panels, and ventilation systems. Its toughness leads to its use for crates, especially for beer and soft drinks. It is also widely used as a fibre in the production of carpet pile and backing, for sacks and cloth, and for ropes and twine.

Nylon: Nylon is strong and tough with good abrasion-resistance. It is also resistant to chemicals, fuels, and oils and is often used to replace metals, particularly die-castings. Nylon has excellent wear properties and in many instances can be used for applications without lubrication.

It is widely used for light engineering components such as gears, especially where silent running is required; bushes; bearings; door-lock components; hinges; and latches.

Nylon is non-toxic and is used in film form for food packaging and in the packaging of surgical instruments where sterilisation is required.

Polytetrafluoroethylene (PTFE): One of the most expensive plastics, PTFE has exceptional resistance to chemical attack, a very low coefficient of friction, excellent electrical insulating properties, and can be used continuously at temperatures up to 250°C.

Limitations of PTFE are its poor abrasion-resistance, high coefficient of expansion, and poor thermal conductivity. By adding filler materials such as glass fibre, graphite, and finely powdered metal, these characteristics can be modified and improved.

The major applications are in chemical engineering, where its exceptional resistance to chemical attack makes it useful for handling highly corrosive chemicals and solvents; in electrical engineering as an insulating material; and in mechanical engineering, where its low frictional properties make it ideal for bearings.

14.3 Working in plastics

Various techniques are available within a conventional workshop for producing shapes in plastics materials without the need of expensive equipment.

The ease with which plastics can be moulded and formed to shape is one of the major advantages of this range of materials. It is not always practicable to produce the required shape, often for economical reasons; for example, a small quantity would not justify the high cost of moulding equipment. Small-scale casting or forming techniques may have to be used.

The moulding techniques used may not give the required accuracy. In this case a machining operation may have to be carried out.

Sheet plastics material may require to be fabricated, in which case sheets can be joined by welding.

14.4 Welding

Hot-gas welding

This is a welding process which consists of heating and softening the two surfaces to be joined and a filler rod, usually of the same material, until complete fusion takes place. It is a similar process to welding metal, except that a naked flame is not used as this would burn the plastics material. Instead a stream of hot gas is used, directed from a special welding torch. The heat source may be electric or gas, at a temperature around 300°C.

The surfaces to be welded have to be prepared to accept the filler rod. For butt joints the edges are chamfered to an included angle of about 60°. For fillet joints an angle of 45° is used, as shown in Fig. 14.1. The surfaces to be welded must be clean and free from grease. Filler rods are generally circular in section and for small work are usually 3 mm diameter.

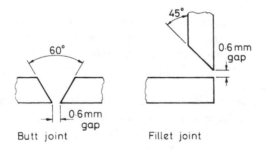

Figure 14.1 Butt- and fillet-weld preparation

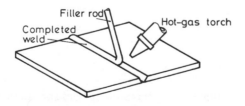

Figure 14.2 Hot gas welding

The surfaces to be joined should be clamped together. The hot air is then directed at the surfaces and the filler rod, which is pressed into the joint as the area becomes tacky. The downward pressure of the filler rod makes the weld, which fuses and solidifies as welding proceeds, Fig. 14.2. Depending on the thickness of the material, more than one weld run may be necessary.

This method is used with good results to weld rigid PVC, polypropylene, and polyethylene (polythene) sheet to fabricate tanks, vessels, pipes for all types of fluids, and ducting.

Heat sealing

This method of welding uses a heated metal strip or bar at a temperature of between 180°C and 230°C. The heated strip is applied under pressure to the surfaces to be welded. To prevent the plastics being sealed sticking to the heated bars, a material such as PTFE is placed between them and the bar. This method is used on nylon and with polyethylene sheet, e.g. in the manufacture of polyethene bags.

Solvent welding

Solvents can be used to soften thermoplastic materials, which, if placed together, will then completely fuse when the solvent evaporates. The main disadvantage of solvent welding is the risk of some of the solvent reaching surfaces other than those being joined and leaving a mark.

The main uses of solvents are with polystyrene used in the manufacture of toys and model kits and with acrylics (Perspex) in the manufacture of display signs, ornaments, and models.

Care must be taken when using solvents, as many are flammable and give off toxic vapours.

14.5 Machining

Machining of plastics is carried out when the number of workpieces to be produced is small and to purchase expensive moulding equipment would be uneconomic. Alternatively, where accuracy greater than can be obtained from the moulding technique is required or where features such as tapped holes cannot be included, machining is essential.

Most plastics materials can be machined using metal-working tools and machines. Since plastics materials have a low thermal conductivity and a high coefficient of expansion, heat produced in cutting must be kept at an absolute minimum. To minimise this heat, it is advisable to use a cutting fluid and to grind all tools with larger clearance angles than are necessary when cutting metal. When machining plastics materials, take light cuts and use high cutting speeds with low feed rates.

The large range of plastics materials available makes it difficult to be specific, and the following is offered only as a general guide to the common machining techniques.

Sawing

In general with all soft materials, coarse-tooth hacksaw blades should be used to prevent clogging the teeth. With brittle materials such as acrylics (Perspex), it may be necessary to use a finer-tooth saw to avoid splintering the edges.

Drilling

Standard high-speed drills are satisfactory for use with plastics but must be cleared frequently to remove swarf. Slow-helix drills (20° helix) reduce the effect of swarf clogging in the flutes and, with a drill point of around 90°, give a better finish on break-through with the softer plastics. All drills should have an increased point clearance of 15° to 20°.

In drilling thin plastics sheet, a point angle as great as 150° is used on the larger diameter drills, so that the point is still in contact with the material when the drill starts cutting its full diameter. Alternatively, the sheet can be clamped to a piece of waste material as with thin sheet metal.

Use cutting speeds of around 40m/min with a feed rate of 0.1 mm/rev. Problems associated with drilling plastics materials were discussed in Section 7.7.

Reaming

Helical-flute reamers should always be used. The reamer must be sharp, otherwise the material tends to be pushed away rather than cut.

Turning

High-speed-steel cutting tools in a centre lathe can be used to turn plastics materials. Clearance angles should be increased to around 20°. A rake angle of 0° can be used on the more brittle plastics, while a 15° rake used on softer plastics aids the flow of material over the tool face. Cutting edges must be kept sharp. Cutting speeds of 150m/min and higher are used, with feed rates between 0.1 and 0.25 mm/rev.

Milling

Milling plastics materials can be carried out using high-speed-steel cutters on a standard milling machine. The cutters should be kept sharp and in good condition.

To avoid distortion, the less rigid plastics materials should be supported over their complete area, due to the high cutting forces acting. Cutting speeds and feed rates similar to those for turning can be used.

Tapping and threading

Holes drilled or moulded in plastics materials can be tapped using high-speed-steel ground-thread taps. With softer plastics materials there is a tendency for the material to be pushed away rather than cut, and this may necessitate the use of special taps about 0.05 mm to 0.13 mm oversize.

Threading can be carried out using single-point tools in a centre lathe with the same angles as for turning. High-speed-steel dies can also be used, but care must be taken to ensure that the thread is being cut and the material is not merely being pushed aside.

Threads cut directly in plastics materials will not withstand high loads and will wear out if screw fasteners are removed and replaced several times. Where high strength and reliability are required, threaded inserts are used, Fig. 14.3. These can be pushed into a predrilled hole, the action of screwing in a fastener causing the knurled outside diameter to bite into the plastics. Threaded inserts may be headed or headless and are available from M2 to M8 thread sizes.

14.6 Heat bending

The forming of thermoplastics can be conveniently carried out by applying heat, usually between 120°C and 170°C, and bending to shape. Care must be taken not to overheat, as permanent damage to the material can result. Provided no permanent

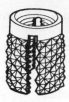

Figure 14.3 Threaded insert for plastics materials

damage has been done, a shaped thermoplastic sheet will return to a flat sheet on the application of further heat.

Simple bending is carried out by locally heating along the bend line, from both sides, until the material is pliable, using a strip heater. A strip heater can easily be constructed using a heating element inside a box structure, with the top made from a heat-resisting material. The top has a 5 mm wide slot along its centre, through which the heat passes, Fig. 14.4. When the material is pliable, it can be located in a former and bent to the required angle, e.g. in making a splash guard for a lathe, Fig. 14.5. The material can be removed from the former when the temperature drops to about 60°C. Formers can be simply made from any convenient material such as wood.

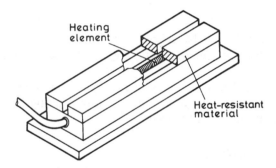

Figure 14.4 Strip heater

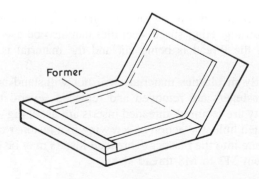

Figure 14.5 Bending a splash guard for a centre lathe

Shapes other than simple bends can be carried out by heating the complete piece of material in an oven. To avoid marking the surface, the material can be placed on a piece of brown paper. The time in the oven depends on the type of material and its thickness, and time must be allowed for the material to reach an even temperature throughout.

Acrylic sheet material is easily worked at 170°C, 3 mm thickness requiring about 20 minutes and 6 mm thickness about 30 minutes in the oven. Again, a simple former can be used to obtain the required shape, e.g. in making a guard for a drilling machine, Fig. 14.6.

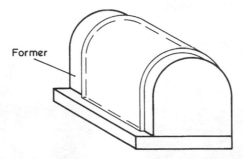

Figure 14.6 Forming a drill guard

14.7 Encapsulation

Encapsulation is the term used when a component, or part of it, is completely enclosed in a plastics material. Encapsulation is carried out for three reasons:

- *protection* – many components, especially small electronic parts, are extremely fragile and must be protected from mechanical damage, e.g. through being dropped or knocked or from vibration;
- *insulation* – many components are required to function over a range of temperatures and therefore require thermal insulation; electrical and electronic components and circuits require insulation from electric current;
- *sealing* – many components and circuits will not work efficiently or may cease to function if they corrode or come into contact with moisture.

The materials used in encapsulation are thermosetting plastics based on phenolic, epoxy, polyester, and silicone resins and compounds. These may be transparent or coloured, depending upon the application.

The silicones are particularly suited to encapsulation, as they can be produced as fluids, gums, and resins and, with the addition of fillers, can be further modified to produce rubbers, greases, and compounds. The silicones are unaffected by temperatures from –50°C to + 200°C and in some instances as high as + 300°C. They provide excellent thermal and electrical insulation and are highly resistant to moisture and oxidation.

The process of encapsulation can be carried out by one of four methods: casting, potting, moulding, or coating.

Casting
This method is used to encapsulate metallurgical specimens to provide a permanent record. The specimen is placed upside down in a mould, and a resin/catalyst mixture is poured round it until the specimen is completely covered. The resin is then left to harden, or cure, the cure time depending on the amount of catalyst used and the surrounding temperature.

When the resin is cured, the casting is removed from the mould. The specimen is then polished using progressively finer grades of wet-and-dry abrasive paper until all scratch marks are removed and the surface is highly polished. The metallurgical detail of the surface is shown up by etching with an acid solution prior to examination under a microscope.

Figure 14.7 shows parts of a gear encapsulated in a phenolic compound and two items in a transparent epoxy resin.

Figure 14.7 Encapsulated metallurgical specimens

Potting

This method is similar to casting except that the encapsulated component is not removed from the mould or pot after curing. The component is placed in the pot and the resin/catalyst mixture is poured round it until the pot is full. The resin is allowed to cure and the process is complete.

The transformer shown in Fig. 14.8 has been potted in a transparent resin to provide a seal against moisture.

Figure 14.8 Transformer before and after encapsulation by potting

Moulding

This method uses a moulding compound which may be in the form of powder or preformed pellets. The compound is forced into a heated mould in which the

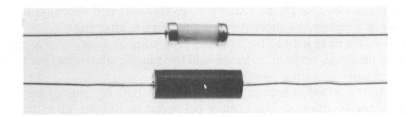

Figure 14.9 Resistor before and after encapsulation by moulding

component has been placed, flows round the component, and cures due to the heat. This process is called transfer moulding and can be used to encapsulate transistors, integrated circuits, resistors, diodes, rectifiers, and other electronic components.

The resistor shown in Fig. 14.9 is encapsulated by this method in a silicon moulding compound to give service at high temperatures.

Coating

Some components, especially electronic circuits, require to be coated to prevent moisture from collecting and so eliminate arcing or tracking. This can be done by spraying, brushing, or dipping, giving an even coat over the complete circuit and its components. The silicones are used for this type of encapsulation.

14.8 Plastics moulding processes

As already stated, the shaping of plastics material is achieved by the application of heat and pressure. There are a great many ways in which this can be done, depending on the nature of the polymer, the type and size of product, and the quantity and dimensional accuracy required. The methods to be described here are moulding by the compression, transfer, and injection processes.

Before dealing with the moulding process itself, however, it is necessary to consider the polymer material, which is usually unsuitable for moulding until mixed with other ingredients called additives.

Forms of supply

Plastics materials for use in moulding are normally in the form of powders or small chips known as granules, or as preforms.

In their pure unmodified state, polymer materials resulting from industrial polymerisation are in most cases unsuitable for processing into finished articles. Before being moulded, they have to be mixed with other ingredients – known as additives – in order to modify or eliminate undesirable properties and to develop their useful characteristics.

Additives can be incorporated into the monomer before polymerisation, during the polymerisation reaction, or with the polymer itself. Some additives modify the properties of the polymer by physical means, while some achieve their effect by chemical reactions and are used not only to influence the properties of the finished product but also to improve processing characteristics. Some of the more common additives are as follows.

- *Plasticisers* – added particularly to PVC to give greater flexibility and make it easier to form.
- *Stabilisers* – added, again particularly to PVC, to prevent decomposition of the polymer at temperatures encountered during normal processing.
- *Lubricants* – widely used to facilitate the processing of a variety of polymers, by reducing the forces between molecules and by reducing adhesion of the polymer to hot metal surfaces during processing.
- *Antioxidants* – added to prevent oxidative degradation, i.e. gradual breakdown in the presence of oxygen, which most polymers are subject to at the elevated temperatures necessary for processing and at atmospheric temperatures over a period of time.
- *Fillers* – added to improve physical properties or in some cases to give a cheaper product by acting as extenders. Fillers used include wood flour, cork dust, asbestos, carbon black, chalk, and chopped glass fibre.
- *Ultra-violet absorbers* – added to protect polymers from the adverse affect of exposure to ultra-violet radiation, the main source of which is sunlight.
- *Flame retardants* – added as most polymers are flammable to a greater or lesser extent.
- *Colourants* – added for a number of reasons:
    i) to give a greater product appeal and make the product more saleable;
    ii) as a means of identification (e.g. cable insulation);
    iii) to make the product more readily visible (e.g. garments for roadmen and motor cyclists);
    iv) to simulate a natural or traditional product (e.g. leather luggage).

Colourants may be added as dyes or, more commonly, as pigments and are available in a vast range of colours.

Moulding processes

Metals have a definite melting point and in general tend to be free-flowing in a molten state. Polymers, on the other hand, have no definite melting point but are softened by the application of heat, which renders them 'plastic'. In this state they may be considered as very viscous fluids and, as a result, high pressures are required for moulding.

The viscosity of a polymer is reduced by the application of heat, but there is an upper temperature limit at which the polymer begins to break down in some way. This breakdown is known as degradation. All polymers are bad conductors of heat and are therefore susceptible to over-heating. If a polymer is subjected to excessive temperature or to prolonged periods in the mould, degradation will occur.

There is also a lower limit of temperature below which the polymer will not be soft enough to flow into the mould. The temperature for moulding must be between the upper and lower limits and will directly affect the viscosity of the polymer.

All the moulding processes require three stages:

1. application of heat to soften the moulding material;
2. forming to the required shape in a mould;
3. removal of heat.

The three moulding techniques to be discussed essentially differ only in the way the moulding material is heated and delivered to the mould.

Compression moulding: Compression moulding is used for thermosetting plastics. The process is carried out in a hydraulic press with heated platens. The two halves of the moulding tool consist of a male and a female die, to give a cavity of the required finished shape of the product, and are attached to the platens of the press, Fig. 14.10. Depending on the size and shape of the product and on the quantity required, the moulding tool may contain a single cavity or a number of cavities, when it is known as a multi-cavity mould. The mould cavity is designed with an allowance for shrinkage of the moulding material and a draft angle of at least 1° to allow for the escape of gases and easy removal of the product after moulding. Moulding tools are manufactured from tool steel, hardened and tempered to give strength, toughness, and a good hard-wearing surface. The male and female dies are highly polished and, once the tool has been proved, these surfaces are chromium-plated (around 0.005 mm thickness) to give a high surface gloss to the product and to facilitate its removal and protect the surfaces from the corrosive effects of the moulding material.

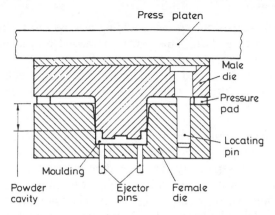

Figure 14.10 Compression mould

Moulding materials are normally in the form of loose powder or granules. These materials have a high bulk factor, i.e. the volume of the loose material is much greater than that of the finished product. The bulk factor is around 2.5:1 for granulated materials and can be as high as 4:1 for fine powder. To allow for this, a powder cavity is built into the female die attached to the bottom press platen. To prevent an excess of loose material being loaded into the tool, each charge is weighed, either on scales or by some automatic method.

Alternatively, loose powder material can be compressed to form a small pellet, or preform, of a size and shape to suit the mould cavity. This is done cold, so that no curing takes place, and is carried out in a special preforming or pelleting machine. The preform is easily handled and gives a consistent mass of charge. Depending on the material, these preforms may be preheated to around 85°C in a high-frequency oven – this reduces the cycle time, since the preform is partially heated before it is loaded in the tool and requires less pressure in moulding.

The typical sequence of operations for compression moulding is as follows:

1. Load moulding material – as loose powder, granules, or a heated preform – into the heated die cavity. Moulding temperatures vary, e.g. between 135°C and 155°C for urea powders and between 140°C and 160°C for melamine materials.
2. Close the split mould between the press platens. The pressure is around 30 N/mm^2 to 60 N/mm^2 (lower for preheated pellets). The combined effect of the moulding

temperature and pressure causes the moulding material to soften and flow into the mould cavity. Further exposure to the moulding temperature causes the irreversible chemical reaction of cross-linking or curing. The curing time depends on the wall thickness, mass, moulding material, and moulding temperature; for example, a 3 mm section of urea will cure in around 30 s at 145°C and the same section in melamine in up to 2 min at 150°C.

3. Open the split mould.
4. Eject the product from the mould. This may be done by hand or automatically, according to the complexity of the tool or product. Since the material is thermosetting and has cured, there is no need to wait for the moulding to cool and therefore it can be removed immediately while it is still hot.
5. Blow out the tool to remove any particles left behind by the previous moulding.
6. Lubricate the tool to assist the release of the next moulding.
7. Repeat the process.

Any material which escapes during the moulding process results in a feather edge on the moulding known as a flash. This can be removed by the operator during the curing cycle of the next moulding.

Compression moulding is used to produce a wide range of products, including electrical and domestic fittings, toilet seats and covers, bottle tops and various closures, and tableware.

Transfer moulding: Transfer moulding is used for thermosetting plastics. This process is similar to compression moulding except that the plasticising and moulding functions are carried out separately. The moulding material is heated until plastic in a transfer pot from which it is pushed by a plunger through a series of runners into the heated split mould where it cures, Fig. 14.11. The two halves of the split mould are attached to the heated platens of a hydraulic press in the same way as for compression moulding.

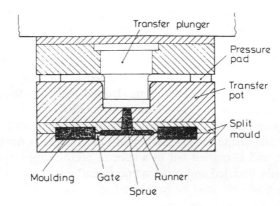

Figure 14.11 Transfer mould

The split mould is closed and the moulding material in the form of powder, granules, or a heated preform is placed in the transfer pot and pressure is applied. The pressure on the area of the transfer pot is greater than that in the compression-moulding process but, as the moulding material is plastic when it enters the mould cavity, the pressure within the cavity is much less. As a result of this, the process is suited to the production of parts incorporating small metal inserts. Intricate parts and those having variations of section thickness can be produced to advantage by this method. Cure times are less and greater accuracy is achieved than with compression moulding.

The main limitation of the process is the loss of material in the sprue, runners, and gates – as thermosetting materials cure during moulding, this cannot be reused. Moulding tools are usually more complex and therefore more costly than compression-moulding tools.

Typical products produced by this process are motor-car distributor caps and domestic electric plugs.

Injection moulding: Injection moulding may be used for either thermoplastics or thermosetting materials, but is most widely used for thermoplastics.

The major advantages of this moulding process are its high production rate, high degree of dimensional accuracy, and its suitability for a wide range of products.

The moulding material is fed by gravity from a hopper to a cylindrical heating chamber where it is rendered plastic and then injected into a closed mould under pressure. The moulding solidifies in the mould. On solidification, the mould is opened and the moulding is ejected.

The type of injection-moulding machine most widely used is the horizontal type, Fig. 14.12. A hopper at the opposite end of the machine from the mould is charged with moulding material in the form of powder or granules which is fed by gravity to the heating chamber. Electric band heaters are attached to the outer casing of the heating chamber, inside which is an extruder-type screw similar to that of a domestic mincing machine. As the screw rotates, it carries the material to the front of the heating chamber. The band heaters and the frictional forces developed in the material by the rotating screw result in the material becoming plastic as it passes along.

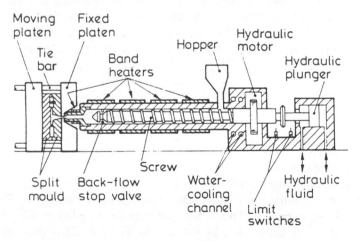

Figure 14.12 Injection-moulding machine

When the plasticised material builds up in front of the screw, the screw moves axially backwards and its rotation is stopped. The amount of material to be injected into the mould – known as the shot size – is controlled by stopping the screw rotation at a predetermined position. At this stage, the mould, filled by the previous shot, is opened and the moulding is ejected. The mould is then closed and the stationary screw is moved axially towards the mould, pushing the plasticised material into the mould cavity under pressure. The screw then rotates, feeding more moulding material along the heating chamber to become plasticised, the material being continuously replaced from the hopper. The screw then moves axially backwards due to the build-up, and the sequence recommences. The heating-chamber temperature varies between 120°C and 260°C, depending on the type of moulding material and the shot size.

Moulding tools are of the same materials and finish as described for compression-moulding tools but without, of course, the need for a powder cavity. They are generally more expensive than compression-moulding tools but, because of the higher production rates of injection moulding, it is possible to use smaller, i.e. single-cavity, and hence cheaper tools and maintain the same number of mouldings per hour as would be possible with a compression press.

The use of thermoplastics material requires that the mould be maintained at a constant temperature – usually around 75°C to 95°C – to cool and solidify the material within the mould before the moulding can be ejected. This is achieved by circulating water through the mould and makes the process much faster than compression moulding. Although material is used in the sprue and runners, material wastage is low since it can be reused.

Injection moulding of thermosetting materials is achieved in the same way as for thermoplastic materials except that the temperatures used are more critical. The temperature of the heating chamber is important – to avoid the moulding material curing before it enters and fills the mould cavity – and is between 95°C and 105°C for urea materials and between 100°C and 110°C for melamine materials. Moulding-tool temperature is also important, to ensure correct curing of the moulding material. Tool temperatures, between 135°C and 145°C for urea materials and between 145°C and 155°C for melamine materials are suitable for sections over 3 mm and are increased by 10°C for sections below 3 mm. As the moulding material enters the mould at very near to the curing temperature, cycle times are low.

The range of injection-moulded components is vast and includes toys, e.g. model kits; houseware, e.g. buckets, bowls, and washing-machine parts; and car components.

Inserts

The primary purpose of inserts is to strengthen relatively weak plastics materials in order to facilitate the joining of mouldings or the mounting of other parts to the moulding. Inserts may also be used for electrical purposes, to provide a conductor in an otherwise insulating material.

Inserts are usually made from steel or brass and may be moulded-in during the moulding process or introduced into a plain hole produced during moulding or by a subsequent drilling operation. This latter type may be installed in the plain hole by using heat (thermal or ultrasonic), by pushing in, or by a self-tapping arrangement. Installation using heat is normally confined to the thermoplastic materials. Whichever type of insert is used, it is essential that an adequate means of anchorage is provided to prevent movement or pulling out of the insert in use.

Use of inserts for electrical purposes can be illustrated by a motor-vehicle distributor. The distributor cap has moulded-in inserts as shown in the cut-away view in Fig. 14.13. Examples of inserts for fixing or mounting purposes, again moulded-in, can be seen in the 13 A plug top and the electrical connection box shown in Fig. 14.14.

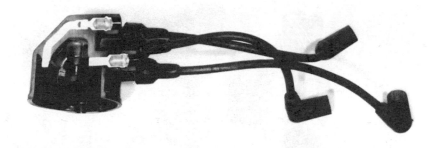

Figure 14.13 Inserts in distributor cap

Figure 14.14 Inserts in electrical components

A wide variety of threaded inserts, both internal and external, are available with a variety of exterior arrangements for inserting and anchoring them in a premoulded or predrilled hole. Figure 14.15 shows internally threaded inserts with a 'barb' form of exterior. These may be installed by pushing in or, in the case in the illustration, by ultrasonic installation where ultrasonic vibration causes instant localised melting in the clear acrylic knob. Figure 14.16 shows externally threaded inserts – the larger thread form is tapered and contains a slot to give a self-tapping action. These self-tapping inserts are installed by hand or power tools to give a finished product as shown.

Advantages and limitations of moulding processes

Advantages and limitations of the moulding processes are summarised in Table 14.1.

To select the appropriate moulding process to be used for a given component, a number of factors may be considered, some of which may automatically exclude a particular process. Among these factors, referred to in Table 14.1, are

- the type of material to be moulded;
- the dimensional accuracy required;
- the required output.

For example, a small quantity of mouldings to be made from a thermosetting material and not requiring a high degree of accuracy would favour the compression-moulding

Figure 14.15 Internal-threaded inserts

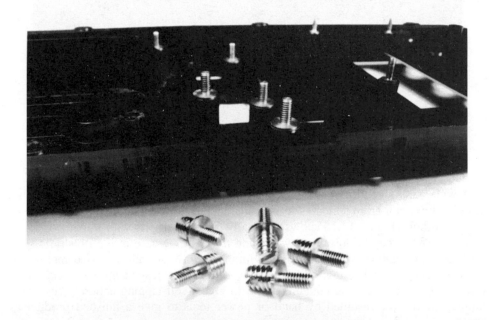

Figure 14.16 External-threaded inserts

process; larger quantities of higher accuracy from a thermosetting material would favour transfer moulding; while high-volume production from thermoplastics material would be carried out using the injection-moulding process.

Safety in plastics moulding

Any machine with moving parts is a potential hazard and as such must be properly guarded. Moulding machines used with plastics are basically presses using high forces to lock the dies in position while the moulding operation is carried out. Adequate guarding must therefore be provided and used to prevent the operator coming in contact with moving parts and so eliminate the possibility of trapped fingers and hands.

Heat is an essential part of any moulding process. Moulding materials have a high heat capacity and in their hot plastic state will stick on contact with the skin and are difficult to remove. Protection can be afforded by the use of suitable industrial gloves. Some of the materials used can cause dermatitis, which can be prevented by the use of gloves.

Harmful gases and vapours are given off by some plastics materials, so the moulding machine must be fitted with adequate extraction equipment.

Table 14.1 Advantages and limitations of moulding processes

	Moulding process		
	Compression	Transfer	Injection
Capital machine cost	Medium. May require preheat ovens and pelleting machines.	Medium	High
Moulding-tool cost	Low	Medium	High
Output rate	Low	Medium	High
Dimensional accuracy	Low	High	High
Moulding material	Thermosets	Thermosets	Mainly thermoplastics
Finishing required	Flash removal	Removal of flash, sprue, and runners	Removal of sprue and runners
Inserts moulded in?	Not usual	Yes	Yes
Waste material	None	Sprue and runners cannot be reused	With thermoplastics, sprue and runners can be reused

15 Primary forming processes

Most metal objects have at some stage in their manufacture been shaped by pouring molten metal into a mould and allowing it to solidify. On solidifying, the object is known as a casting if the shape is such that no further shaping is required – it may only require machining to produce the finished article. Castings are produced by various methods, for example sand casting, die-casting, and investment casting. If, upon solidifying, the object is to be further shaped by rolling, extrusion, drawing, or forging, it is known as an ingot, pig, slab, billet, or bar – depending on the metal and the subsequent shaping process – and these are cast as simple shapes convenient for the particular forming process.

Most metals, with the exception of some precious ones, are found in the form of minerals or ores. The ores are smelted to convert them into metals; for example, iron is obtained from iron ore (haematite), aluminium from bauxite, and copper from copper pyrites.

Iron ore together with other elements is smelted in a blast furnace to give pig iron. According to the type of plant, the molten pig iron is cast as pigs or is transferred to a steel-making process. Cast pigs are refined in a cupola to give cast iron which is cast as notched ingots or bars of relatively small cross-section for ease of remelting in the foundry as required. Pig iron taken in a molten state to the steel-making process is made into steel which is cast as ingots or now more usually as slabs produced, by the continuous casting process for subsequent rolling, drawing, or forging.

Aluminium is extracted from bauxite by an electrolytic process. Commercially pure aluminium is soft and weak and it is alloyed to improve the mechanical properties. Aluminium alloys are available wrought or cast, in sections convenient for subsequent working by rolling, drawing, casting, or extrusion.

Copper is extracted from copper pyrites and is refined by remelting, in a furnace or electrolytically. Copper is alloyed to produce a range of brasses and bronzes. These materials may then be rolled, drawn, cast, or extruded.

15.1 Forms of supply of raw materials

Where the metal has to be remelted, as in the case of casting, it is usual to supply it to the foundry in the form of notched ingots or bars of relatively small cross-section. These can be broken into smaller pieces for ease of handling and loading into small furnaces.

Where the metal is to be subjected to further forming processes – e.g. rolling, extrusion, drawing, or forging – it would be wasteful to remelt, both from an energy viewpoint and from the effect the remelting would have on the physical and mechanical properties. In this case the raw material would be supplied in the form most convenient for the process – i.e. slabs (width greater than three times the thickness) for rolling into sheet; blooms and billets (smaller section) for rolling into bar, sections, etc. and for forging and extrusion. Hot-rolled rod is supplied for cold drawing into rod, tube, and wire.

15.2 Properties of raw materials

Fluidity: This property is a requirement for a metal which is to be cast. The metal must flow freely in a molten state in order to completely fill the mould cavity.

Ductility: A ductile material can be reduced in cross-section without breaking. Ductility is an essential property when drawing, since the material must be capable of flowing through the reduced diameter of the die and at the same time withstand the pulling force. The reduction in cross-sectional area aimed for in a single pass through the die is usually between 25% and 45%.

Malleability: A malleable material can be rolled or hammered permanently into a different shape without fracturing. This property is required when rolling and forging.

Plasticity: This is a similar property to malleability, involving permanent deformation without fracture. This property is required in forging and extrusion, where the metal is rendered plastic, i.e. made more pliable, by the application of heat.

Toughness: A material is tough if it is capable of absorbing a great deal of energy before it fractures. This property is required when forging.

15.3 Sand casting

Casting is the simplest and most direct way of producing a finished shape from metal. Casting shapes from liquid metals can be done by a variety of processes, the simplest of which is sand casting.

For the production of small castings, a method known as box moulding is employed. The box is made up of two frames with lugs at each end into which pins are fitted to ensure accurate alignment when the frames are placed together. The top frame is referred to as the cope and the bottom frame as the drag.

A removable pattern is used to create the required shape of cavity within the mould. The pattern is made in two parts, split at a convenient position for ease of removal from the mould – this position being known as the parting line. The two parts of the pattern are accurately aligned with each other using pins or dowels.

Consider the gear blank shown in Fig. 15.1(a), the pattern for which is shown in Fig. 15.1(b). One half of the pattern is placed on a board. The drag is then placed on the board such that the pattern half is roughly central. Moulding sand is then poured into the drag and is pressed firmly against the pattern. The drag is then filled completely and the sand is firmly packed using a rammer, Fig. 15.2(a). The amount of ramming should be sufficient for the sand to hold together but not enough to prevent the escape of gases produced during pouring of the molten metal. After ramming is complete, the sand is levelled off flush with the edges of the drag. Small vent holes can be made through the sand to within a few millimetres of the pattern, to assist the escape of gases.

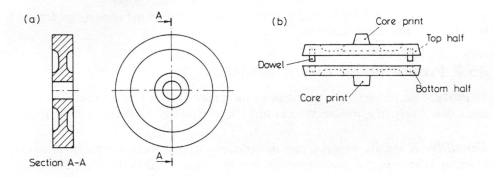

Figure 15.1 Gear blank and pattern

The drag is then turned over and the second half pattern is located by means of the dowels. The upturned surface is then covered with a fine coating of dry 'parting sand', to prevent bonding between the sand in the cope and that in the drag. The cope is accurately positioned on the drag by means of the locating pins. To allow entry for the molten metal, a tapered plug known as a runner pin is placed to one side of the pattern. A second tapered plug known as a riser pin is placed at the opposite side of the pattern, Fig. 15.2(b) – this produces an opening which, when filled with molten metal during pouring, provides a supply of hot metal to compensate for shrinkage as the casting cools. The cope is then filled, rammed, and vented as for the drag. The pins are then removed and the top of the runner hole is enlarged to give a wide opening for pouring the metal.

The cope is then carefully lifted off and turned over. Both halves of the pattern are carefully removed. Small channels known as gates are then cut from the bottom of the runner and riser to enable metal to fill the mould cavity. Any loose sand is blown away to leave a clean cavity in the mould. The core is placed in position in the bottom half of the mould box (the drag) and the top mould box (the cope) is carefully replaced in position with the aid of the locating pins and is clamped to prevent lifting when the metal is poured, Fig. 15.2(c). The core which is necessary to provide a hollow section, in this case the bore of the gear blank, is made separately.

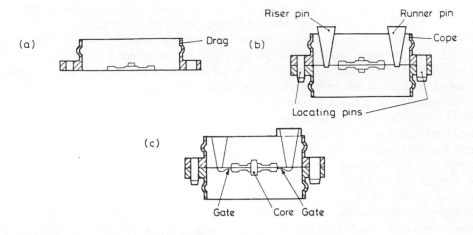

Figure 15.2 Basic steps in sand casting

The mould is now ready for pouring. When the metal has solidified, the mould is broken up to release the casting. The runner and riser are broken off, and the rough edges are removed by fettling (i.e. hand grinding).

Patterns

For the production of small quantities of castings, patterns are made from wood, smoothed, painted, or varnished to give a smooth finish to the casting. Patterns are made larger than the finished part, to allow for shrinkage of the casting when it cools. A special rule, known as a contraction rule, is available to suit different metals – the pattern-maker makes the pattern using measurements from the contraction rule, which automatically gives the correct dimension of pattern with due allowance for shrinkage whatever the size of dimension.

Some surfaces of a casting may require subsequent machining, e.g. surfaces requiring a greater accuracy of size, flatness, or surface finish than can be achieved by casting. Extra metal must be left on these surfaces, and the amount to be left for removal by machining must be allowed for on the appropriate surface of the pattern.

Where cores are to be incorporated in a casting (see below), provision must be made on the pattern to provide a location seating in the mould. These sections added to the pattern are known as core prints.

To allow the pattern to be easily removed from the mould, a small angle or taper known as draft is incorporated on all surfaces perpendicular to the parting line.

Sand

Moulding sand must be permeable, i.e. porous, to allow the escape of gases and steam; strong enough to withstand the mass of molten metal; resist high temperatures; and have a grain size suited to the desired surface of the casting.

Silicon sand is used in moulding, the grains of sand being held together in different ways.

In green-sand moulds the grains are held together by moist clay, and the moisture level has to be carefully controlled in order to produce satisfactory results.

Dry-sand moulds start off in the same way as green-sand moulds but the moisture is driven off by heating after the mould has been made. This makes the mould stronger and is suited to heavier castings.

With CO_2 (carbon dioxide) sand the silica grains are coated with sodium silicate instead of clay. When the mould is made, it is hardened by passing carbon-dioxide gas through it for a short period of time. The sand 'sets' but is easily broken after casting.

Specially prepared facing sand is used next to the pattern to give an improved surface to the casting. The mould can then be filled using a backing sand.

Cores

When a casting is to have a hollow section, a core must be incorporated into the mould. Cores are normally of green sand or dry sand, made separately in a core box, and inserted in the mould after the pattern is removed and before the mould is closed. They are located and supported in the mould in a seating formed by the core prints on the pattern. The core must be strong enough to support itself and withstand the flow of molten metal, and in some cases it may be necessary to reinforce it with wires to give added strength.

The more complex cores are produced from CO_2 sand.

15.4 Rolling

The cast ingots produced by the raw-material producers are of little use for manufacturing processes until they have been formed to a suitable shape, i.e. sheet, plate, strip, bar, sections, etc.

One of the ways in which shapes can be produced in order that manufacturing processes can subsequently be carried out is by rolling. This can be performed as hot rolling or cold rolling – in each case, the metal is worked while in a solid state and is shaped by plastic deformation.

The reasons for working metals in their solid state are firstly to produce shapes which would be difficult or expensive to produce by other methods, e.g. long lengths of sheet, section, rods, etc., and secondly to improve mechanical properties.

The initial stage of converting the ingot to the required shape is by hot rolling. During hot working, the metal is in a plastic state and is readily formed by pressure as it passes through the rolls. Hot rolling has a number of other advantages.

- Most ingots when cast contain many small holes – a condition known as porosity. During hot rolling, these holes are pressed together and eliminated.
- Any impurities contained in the ingot are broken up and dispersed throughout the metal.
- The internal grain structure of the metal is refined, resulting in an improvement of the mechanical properties, e.g. ductility and strength.

Hot rolling does, however, have a number of disadvantages. Due to the high temperatures, the surface oxidises – producing a scale which results in a poor surface finish, making it difficult to maintain dimensional accuracy. Where close dimensional accuracy and good surface finish are not of great importance, e.g. structural shapes for construction work, a descaling operation is carried out and the product is used as-rolled. Alternatively, further work can be carried out by cold rolling.

When metal is cold rolled, greater forces are required, necessitating a large number of stages before reaching the required shape. The strength of the material is greatly improved, but this is accompanied by a decrease in ductility. Depending on the number of stages required in producing the shape, annealing (or softening) may have to be carried out between stages. Besides improving mechanical properties, cold rolling produces a good surface finish with high dimensional accuracy.

In the initial stage of converting it to a more suitable form, the ingot is first rolled into intermediate shapes such as blooms, billets, or slabs. A bloom has a square cross-section with a minimum size of 150 mm square. A billet is smaller than a bloom and may have a square cross-section from 40 mm up to the size of a bloom. A slab is rectangular in cross-section from a minimum width of 250 mm and a minimum thickness of 40 mm. These are then cut into convenient lengths for further hot or cold working.

Most primary hot rolling is carried out in either a two-high reversing mill or a three-high continuous mill.

In the two-high reversing mill, Fig. 15.3, the metal is passed between the rolls in one direction. The rolls are then stopped, closed together by an amount depending on the rate of reduction required, and reversed, taking the material back in the opposite

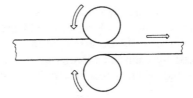

Figure 15.3 Two-high reversing mill

direction. This is repeated, with the rolls closed a little more each time, until the final size of section is reached. At intervals throughout this process, the metal is turned on its side to give a uniform structure throughout. Grooves are provided in the top and bottom rolls to give the various reductions and the final shape where appropriate, Fig. 15.4.

In the three-high continuous mill, Fig. 15.5, the rolls are constantly rotating, the metal being fed between the centre and upper rolls in one direction and between the centre and lower rolls in the other. A platform is positioned such that it can be raised to feed the metal through in one direction or support it coming out from between the rolls in the opposite direction or be lowered to feed the metal back through or to support it coming out from between the rolls in the opposite directions.

In cold rolling, the roll pressures are much greater than in hot rolling – due to the greater resistance of cold metal to reduction. In this case it is usual to use a four-high

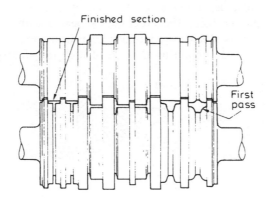

Figure 15.4 Rolls for producing a tee section

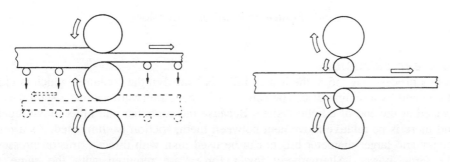

Figure 15.5 Three-high continuous mill   **Figure 15.6** Four-high mill

mill, Fig. 15.6. In this arrangement, two outer rolls of large diameter are used as back-up rolls to support the smaller working rolls and prevent deflection.

When rolling strip, a series of rolls are arranged in line and the strip is produced continuously, being reduced by each set of rolls as it passes through before being wound on to a coil at the end when it reaches its final thickness.

15.5 Extrusion

Extrusion usually has to be a hot-working process, due to the very large reduction which takes place during the forming process. In operation, a circular billet of metal is heated to render it plastic and is placed inside a container. Force is then applied to the end of the billet by a ram which is usually hydraulically operated. This applied force pushes the metal through an opening in a die to emerge as a long bar of the required shape. The extrusion produced has a constant cross-section along its entire length. The die may contain a number of openings, simultaneously producing a number of extrusions.

There are a number of variations of the extrusion process, two common methods being direct extrusion and indirect extrusion.

Direct extrusion

This process, Fig. 15.7, is used for the majority of extruded products. A heated billet is positioned in the container with a dummy block placed behind it. The container is moved forward against a stationary die and the ram pushes the metal through the die. After extrusion, the container is moved back and a shear descends to cut off the butt end of the billet at the die face. The process is repeated with a new billet. As the outside of the billet moves along the container liner during extrusion, high frictional forces have to be overcome, requiring the use of high ram forces.

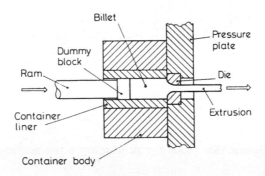

Figure 15.7 Direct extrusion

Indirect extrusion

In this process, Fig. 15.8, the heated billet is loaded in the container which is closed at one end by a sealing disc. The container is moved forward against a stationary die located at the end of a hollow stem. Because the container and billet move together and there is no relative movement between them, friction is eliminated. As a result, longer and larger-diameter billets can be used than with direct-extrusion presses of the same power. Alternatively lower forces are required with the same size billets.

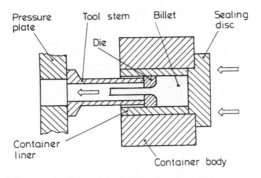

Figure 15.8 Indirect extrusion

The indirect-extrusion process does have limitations, however. Metal flow tends to carry surface impurities into the extruded metal, and the billets have to be machined or chemically cleaned. The size of the extrusion is limited to the inside diameter of the hollow stem, and die changing is more cumbersome than with direct extrusion.

The important features of the extrusion process are:

- the complexity of shape possible is practically unlimited, and finished products can be produced directly, Fig. 15.9;
- a good surface finish can be maintained;
- good dimensional accuracy can be obtained;
- large reductions in cross-sectional area can be achieved;
- the metal is in compression during the process, so relatively brittle materials can be extruded;
- the mechanical properties of the material are improved.

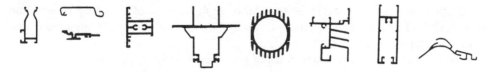

Figure 15.9 Examples of extruded shapes

The extrusion process is, however, limited to products which have a constant cross-section. Any holes, slots, etc. not parallel to the longitudinal axis have to be machined. Due to extrusion-press power capabilities, the size of shape which can be produced is limited. The process is normally limited to long runs, due to die costs, but short runs can be economical with simple die shapes. Typical materials used are copper and aluminium and their alloys.

15.6 Drawing

The primary process of drawing is a cold-working process, i.e. carried out at room temperature. It is mainly used in the production of wire, rod, and bar.

Wire is made by cold drawing a previously hot-rolled rod through one or more dies, Fig. 15.10, to decrease its size and improve the physical properties. The hot-rolled rod – usually around 10 mm in diameter – is first cleaned in an acid bath to

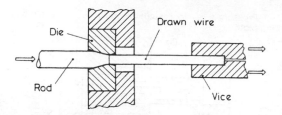

Figure 15.10 Wire drawing

remove scale, a process known as pickling. This ensures a good finish on the final drawn wire. The rod is then washed with water to remove and neutralise the acid. The end of the rod is then pointed so that it can be passed through the hole in the die and be gripped in a vice attached to the drawing machine. The rod is then pulled through the die to give the necessary reduction in section.

In continuous wire drawing, the wire is pulled through a series of progressively smaller dies until the final-size section is reached.

The strength of the material will limit the force which can be applied in pulling the wire through the die, while the ductility of the material limits the amount of reduction possible through each die. Typical materials used are steel, copper, aluminium, and their alloys.

15.7 Forging

Forging is a hot-working process, heat being necessary to render the metal plastic in order that it may be more easily shaped. The oldest form of forging is hand forging as carried out by the blacksmith. Hand tools are used to manipulate the hot metal to give changes in section and changes in shape by bending, twisting, etc. Due to the hand operation, it is not possible to achieve high degrees of accuracy or extreme complexity of shape. This method is limited to one-off or small quantities and requires a high degree of skill.

When forgings are large, some form of power is employed. Steam or compressed-air hammers or a forging press is used, the process being known as open-sided forging. In this process, the hot metal is manipulated to the required shape by squeezing it between a vertically moving die and a stationary die attached to the anvil, Fig. 15.11. This method of forging is carried out under the direction of a forge-master who directs the various stages of turning and moving along the length until the finished shape and size are obtained. Again, great skill is required. This method is used to produce large forgings such as propeller shafts for ships.

When large quantities of accurately shaped products are required, these are produced by a process known as closed-die forging or drop forging. With this method, the hot metal is placed between two halves of a die each containing a cavity such that when the metal is squeezed into the cavity a completed forging of the required shape is produced, Fig. 15.12. The metal is subjected to repeated blows, usually from a mechanical press, to ensure proper flow of the metal to fill the die cavity. A number of stages through a series of dies may be required, each stage changing the shape gradually until the final shape is obtained. The number of stages required depends on the size and shape of the part, how the metal flows, and the degree of accuracy required.

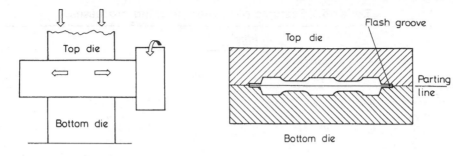

Figure 15.11 Open-die forging   **Figure 15.12** Closed-die forging

One of the two halves of the die is attached to the moving part of the press, the other to the anvil. The cavity contained in the die is designed such that the parting line enables the finished forging to be removed and incorporates a draft in the direction of die movement in the same way as do the pattern and mould in sand casting. The size of cavity also allows for additional material on faces requiring subsequent machining.

Since it is impossible to judge the exact volume required to just fill the die cavity, extra metal is allowed for and this is squeezed out between the two die halves as they close. This results in a thin projection of excess metal on the forging at the parting line, known as a flash. This flash is removed after forging by a trimming operation in which the forging is pushed through a correct-shape opening in a die mounted in a press.

Forging is used in the production of parts which have to withstand heavy or unpredictable loads, such as levers, cams, gears, connecting rods, and axle shafts. Mechanical properties are improved by forging, as a result of the flow of metal being controlled so that the direction of grain flow increases strength. Figure 15.13 shows the difference in the grain flow between a shaft with a flange which has been forged up, Fig. 15.13(a), and one machined from a solid rolled bar, Fig. 15.13(b). Any form machined on the flange, such as gear teeth, would be much weaker when machined from solid bar. The structure of the material is refined due to the hot working, and the density is increased due to compression forces during the forging process.

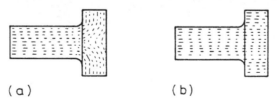

(a)   (b)

Figure 15.13 Direction of grain flow

The process of drop forging is normally restricted to larger batch quantities, due to die costs.

15.8 Selection of a primary process

A number of factors have to be considered before a choice of process for a given component can be made. For example, consideration would have to be given to the type of material to be used, the mechanical properties required, shape, accuracy, degree of surface finish, and the quantity to be produced. Some of these factors are

Table 15.1 Features of primary forming processes

| | Sand casting | Rolling | | Extrusion | Drawing | Forging |
		Hot	Cold			
Improved mechanical properties		✔	✔	✔	✔	✔
Three-dimensional shape variation	✔					✔
Constant cross-section		✔	✔	✔	✔	
Large quantities		✔	✔	✔	✔	✔
Good surface finish			✔	✔	✔	
High dimensional accuracy			✔	✔	✔	
High tool costs				✔		✔

shown in Table 15.1 together with the appropriate primary processes, as an aid to the correct choice. For example, the requirement of variation of shape in three dimensions would eliminate rolling, extrusion, and drawing; a high degree of accuracy would eliminate sand casting and forging; and so on.

16 Presswork

The term 'presswork' is used here to describe the process in which force is applied to sheet metal with the result that the metal is cut, i.e. in blanking and piercing, or is formed to a different shape, i.e. in bending.

The pressworking process is carried out by placing the sheet metal between a punch and die mounted in a press. The punch is attached to the moving part – the slide or ram – which applies the necessary force at each stroke. The die, correctly aligned with the punch, is attached to the fixed part or bedplate of the press.

The press used may be manually operated by hand or by foot and used for light work or it may be power operated, usually by mechanical or hydraulic means, and capable of high rates of production.

The time to produce one component is the time necessary for one stroke of the press slide plus load/unload time or time for feeding the material. Using a power press, this total time may be less than one second.

It is possible to carry out a wide range of operations in a press, and these include blanking, piercing, and bending.

Blanking is the production of an external shape, e.g. the outside diameter of a washer.

Piercing is the production of an internal shape, e.g. the hole in a washer.

Bending – in this case simple bending – is confined to a straight bend across the metal sheet in one plane only.

16.1 Presses

The fly press

Blanking, piercing, and bending of light work where the required force is small and the production rate is low may be carried out on a fly press.

A hand-operated bench-type fly press is shown in Fig. 16.1. The body is a C-shaped casting of rigid proportions designed to resist the forces acting during the pressworking operation. The C shape gives an adequate throat depth to accommodate a range of work sizes. (The throat depth is the distance from the centre of the slide to the inside face of the body casting.) The bottom part of the casting forms the bed to which tooling is attached. The top part of the casting is threaded to accept the multi-start square-threaded screw which carries the handle at its top end and the slide at its bottom end. An adjustable threaded collar is fitted at the top end of the screw and can be locked at a required position to avoid overtravel of the screw during operation. The slide contains a hole and a clamping screw to locate and secure the punch. The horizontal portion of the handle is fitted with ball masses which produce a flywheel effect when larger forces are required. These masses are fitted on spikes, and one or both may be removed when smaller forces are required.

In operation, the vertical handle is grasped and the handle is partially rotated. This provides, through the multi-start thread, a vertical movement of the slide. A punch

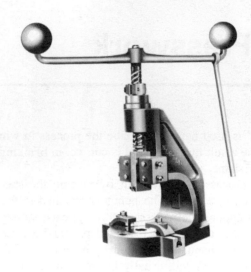

Figure 16.1 Fly press

and die fitted in the slide and on the bed, and correctly aligned with each other, are used to carry out the required press-working operation.

The fly press can be set up easily and quickly for a range of blanking, piercing, and bending operations, and the manual operation gives a greater degree of sensitivity than is often possible with power presses. This type of press may also be used for operations such as pressing dowels and drill bushes into various items of tooling. Due to the manual operation, its production rates are low.

Power presses

Power presses are used where high rates of production are required. A power press may be identified by the design of the frame and its capacity – i.e. the maximum force capable of being delivered at the work, e.g. 500 kN ('50 tons'). The source of power may be mechanical or hydraulic. Different types of press are available in a wide range of capacities, the choice depending on the type of operation, the force required for the operation, and the size and type of tooling used.

One of the main types of power press is the open-fronted or gap-frame type. This may be rigid or inclinable. The inclinable feature permits finished work to drop out the back by gravity. A mechanical open-fronted rigid press is shown in Fig. 16.2. The model shown has a capacity of 1000 kN ('100 tons') and operates at 60 strokes per minutes (it is shown without guards for clarity). The open nature of this design gives good accessibility of the tools and allows the press to be operated from either side or from the front.

The limitation of the open-fronted press is the force which can be applied. High forces have a tendency to flex the frame and so open the gap between the slide and the bedplate. This flexing of the frame can be overcome on the large-capacity presses by fitting tie-rods between the bedplate and the top of the frame as shown in Fig. 16.2.

As the capacity of a press is increased, it becomes necessary to increase the strength and rigidity of its frame – for the reason already outlined. This is achieved in the other main type of power press – the straight-sided or column type – since the

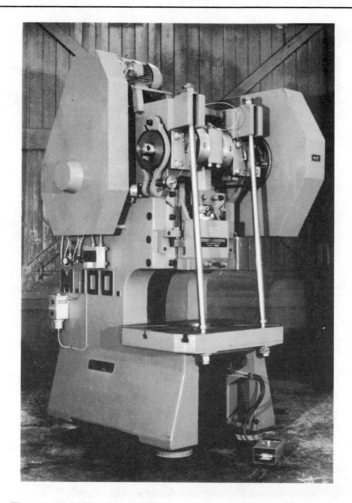

Figure 16.2 Mechanical open-fronted rigid power press

large forces are taken up in a vertical direction by the side frames. A mechanical straight-sided press of 3000 kN ('300 tons') capacity is shown in Fig. 16.3.

Straight-sided presses sacrifice adaptability and accessibility to gain frame rigidity and are best suited to work on heavy-gauge metals and on large surfaces. Since the sides are closed by the side frames and are open at the front and the back, these presses are limited to operation from the front only.

Mechanical presses: Mechanical power presses derive their energy for operation from a constantly rotating flywheel driven by an electric motor. The flywheel is connected to a crankshaft through a clutch which can be set for continuous or single stroking. In the single-stroke mode, the clutch is automatically disengaged at the end of each stroke and the press will not restart until activated by the operator. A connecting rod is attached at one end to the crankshaft and at its other end to the slide. Adjustment is provided to alter the position of the slide, and in some presses the length of stroke can also be adjusted by means of an eccentric on the crankshaft. A brake is fitted to bring the crankshaft to rest at the correct position.

With mechanical presses, the maximum force is available at the bottom of the slide stroke. In blanking and piercing operations, the work is done very near to the bottom

Figure 16.3 Mechanical straight-sided power press

of the stroke. However, where a part is blanked and then taken further down the stroke, e.g. to form a bend, less force will be available for the blanking operation which will have been carried out some distance above the bottom of the stroke. For example, a 500 kN press with a stroke of 120 mm will exert a force of only 120 kN half-way down its stroke.

Hydraulic presses: Hydraulic power presses derive their power from high-pressure hydraulic pumps which operate the ram. The load applied is completely independent of the length and stroke, i.e. full load can be applied at any point in the stroke. The applied load is controlled by a relief valve which gives automatic protection against overload. Again by means of valves, the ram can be made to approach the work rapidly and then be shifted to a lower speed before contacting the work, thus prolonging the life of the tool but still giving fast operating speeds. Rapid ram reversal can also be controlled. Switches are incorporated to determine the positions at which these controls become effective, thereby increasing productivity by making tool setting faster and by keeping the actual working stroke to a minimum.

The number of moving parts are few and these are fully lubricated in a flow of pressurised oil, leading to lower maintenance costs. Fewer moving parts and the absence of a flywheel reduce the overall noise level of hydraulic presses compared to mechanical presses.

Longer strokes are available than with mechanical presses, giving greater flexibility of tooling heights.

A typical hydraulic press of 100 kN ('10 ton') capacity is shown in Fig. 16.4. This model has a ram advance speed of 475 mm/s, a pressing speed of 34 mm/s, and a return speed of 280 mm/s. The model shown is fitted with a light-screen guard operated by a continuous curtain of infra-red light.

Figure 16.4 Hydraulic straight-sided power press

Safety

Many serious accidents to operators and tool setters have occurred in the use of power presses. Stringent safety requirements must be met in the use of power presses, and these are covered by the Power Presses Regulations and the Provision and Use of Work Equipment Regulations 1992 (PUWER) (see Chapter 1). The regulations include requirements for the thorough examination and testing by a competent person of power-press mechanisms and safety devices after installation, before use, and periodically as a condition of use. A vital factor in the prevention of accidents at power presses is the effective maintenance of both presses and guards in sound working condition.

Power-press mechanisms

Press mechanisms such as the clutch, brake, connecting rod, and flywheel journals, as well as guards and guard mechanisms, must be subjected to systematic and regular thorough examination. Many presswork operations involve feeding and removing workpieces by hand, and every effort must be made to ensure that this can be done safely without the risk that the press will operate inadvertently while the operator's hands are within the tool space. The most obvious way of avoiding accidents to the operator is to design the tools in such a way as to eliminate the need for the operator to place his fingers or hands within the tool space for feeding or removal of work. This can often be done by providing feeding arrangements such that the operator's hands are outside the working area of the tools.

All dangerous parts of the press must be guarded, and the four principal ways of guarding the working area are:

- enclosed tools, where the tools are designed in such a way that there is insufficient space for entry of fingers;
- fixed guards, which prevent fingers or hands reaching into the tool space at any time;
- interlocked guards, which allow access to the tools but prevent the clutch being engaged until the guards are fully closed – the guards cannot be opened until the cycle is complete, the clutch is disengaged, and the crankshaft has stopped;
- automatic guards, which push or pull the hand clear before trapping can occur;
- light-screen guards – operated by a continuous curtain of infra-red light which, if broken, stops the machine.

16.2 Press-tool design

The tools used in presses are punches and dies. The punch is attached to the press slide and is moved into the die, which is fixed to the press bedplate. In blanking and piercing, the punch and die are the shape of the required blank and hole and the metal is sheared by passing the punch through the die. In bending, the punch and die are shaped to the required form of the bend and no cutting takes place. In each case, the punch and die must be in perfect alignment. The complete assembly of punch and die is known as a press tool.

Blanking and piercing

In blanking and piercing operations the work material, placed in the press tool, is cut by a shearing action between the adjacent sharp edges of the punch and the die. As the punch descends on to the material, there is an initial deformation of the surface of the material followed by the start of fracture on both sides, Fig. 16.5. As the tensile strength of the material is reached, fracture progresses and complete failure occurs.

The shape of the side wall produced by this operation is not straight as in machining operations and is shown greatly exaggerated in Fig. 16.6. The exact shape depends on the amount of clearance between the punch and the die. Too large a clearance leads to a large angle of fracture and a large burr, while too small a clearance will result in premature wear of the tools and a risk of tool breakage.

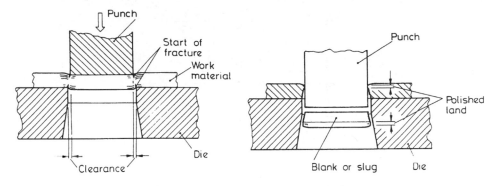

Figure 16.5 Shearing action of punch and die

Figure 16.6 Characteristics of sheared edge

The clearance of the space between the punch and the die is quoted as a percentage of the work-material thickness per side. Establishment of the correct clearance to be used for a given blanking or piercing operation is influenced by the required characteristics of the cut edge and by the thickness and properties of the work material. The values given in Table 16.1 are offered only as a general guide.

Table 16.1 Typical values of clearance for press-tool design

Work material	Clearance per side (% of work-material thickness)
Low-carbon steel	5–7
Aluminium alloys	2–6
Brass:	
annealed	2–3
half hard	3–5
Phosphor bronze	3.5–5
Copper:	
annealed	2–4
half hard	3–5

In blanking and piercing operations, the punch establishes the size of the hole and the die establishes the size of the blank. Therefore in piercing, where an accurate size of hole is required, the punch is made to the required hole size and the clearance is made on the die, Fig. 16.7(a). Conversely, in blanking, where an accurate size of blank is required, the die is made to the required blank size and the clearance is made on the punch, Fig. 16.7(b). On this basis, the material punched through during piercing is scrap and the material left behind in the die during blanking is scrap.

Example 16.1

It is required to produce 50 mm diameter blanks from 2 mm thick low-carbon steel. If a clearance of 6% is chosen, then

    diameter of die = diameter of blank = 50 mm
    clearance per side = 6% of 2 mm = 0.12 mm
    clearance on diameter is therefore 2 × 0.12 mm = 0.24 mm

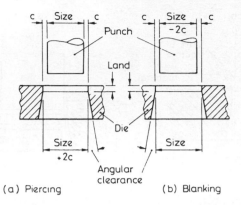

Figure 16.7 Clearance on punch and die for piercing and blanking operations

Thus the diameter of the punch is smaller than the die by this amount; therefore

diameter of punch = 50 mm − 0.24 mm = 49.76 mm

Example 16.2

It is required to punch 20 mm diameter holes in 1.5 mm thick copper. If a clearance of 4% is chosen, then

diameter of punch = diameter of hole = 20 mm
clearance per side = 4% of 1.5 mm = 0.06 mm
clearance on diameter is therefore 2 × 0.06 mm = 0.12 mm

Thus the diameter of the die is larger than the punch by this amount; therefore

diameter of die = 20 mm + 0.12 mm = 20.12 mm

To prevent the blanks or slugs removed by the punch from jamming in the die, it is usual to provide an angular clearance below the cutting edge of the die as shown in Fig. 16.7(a). Thus the pieces can fall through the die, through a hole in the bedplate, into a bin. A land equal in width to approximately twice the metal thickness may be provided, which enables a large number of regrinds to be carried out on the top die face to maintain a sharp cutting edge.

Stripping

As the punch enters the work material during a blanking or piercing operation, it becomes a tight fit in the material. To withdraw the punch without lifting the material along with the punch, it is necessary to provide a method of holding the material while the punch is withdrawn on the upward stroke. This is known as stripping and can be done by providing a fixed stripper or a spring-type stripper.

A fixed stripper, Fig. 16.8, is used when the work material is in the form of strip, fed across the top of the die. The stripper in this case is a flat plate screwed to the top of the die and contains a hole through which the punch passes. When the punch is on the upward return stroke, the plate prevents the material from lifting and strips it off the punch.

When the work material is not in the form of strip and has to be loaded in the tool by hand, the spring-pad type of stripper is used, Fig. 16.9. The stripper pad is set in advance of the punch, holding the material on the die face by means of springs and keeping it flat while the operation is being carried out. Stripper bolts keep the pad in

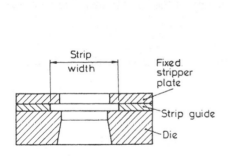

Figure 16.8 Fixed stripper

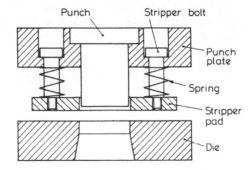

Figure 16.9 Spring-pad stripper

position. When the punch is on the upward return stroke, the pad holds the material against the die and strips it off the punch.

Bending

Bending as described here refers to simple bending, confined to straight bends across the work material in one plane only. In bending operations, no metal cutting takes place: the previously cut material is placed between a punch and a die and force is applied to form the required bend.

Strip or sheet metal should, wherever possible, be bent in a direction across the grain of the material rather than along it. The direction of the grain is produced in the rolling process, and by bending across it there is less tendency for the material to crack. Keeping the bend radius as large as possible will also reduce the tendency for the material to crack.

Bends should not be positioned close to holes, as these can be pulled into an oval shape. It is generally accepted that the distance from the centre of the bend to the edge of a hole should be at least two-and-a-half times the thickness of the work material.

In bending operations, the length of the blank before bending has to be calculated. Any metal which is bent will stretch on the outside of the bends and be compressed on the inside. At some point between the inside and outside faces, the layers remain unaltered in length and this point is known as the neutral axis. For bends of radius more than twice the material thickness, the neutral axis may be assumed to lie at the centre of the material thickness, Fig. 16.10(a). For sharper bends of radius less than twice the material thickness, the neutral axis shifts towards the inside face. In this case, the distance from the neutral axis to the inside face may be assumed to be 0.33 times the material thickness, Fig. 16.10(b).

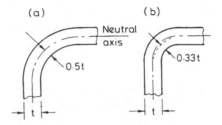

Figure 16.10 Position of neutral axis

The length of the blank is determined by calculating the lengths of the flat portions either side of the radius plus the stretched out length of the bend radius (known as the bend allowance).

☐ *Example 16.3*

Determine the blank length of the right-angled bracket shown in Fig. 16.11(a).

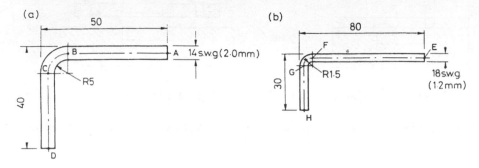

Figure 16.11 Bending examples

Length AB = 50 mm – inside radius – material thickness
           = 50 mm – 5 mm – 2 mm
           = 43 mm
Length CD = 40 mm – 5 mm – 2 mm
           = 33 mm

Since the radius is greater than twice the material thickness t, we can assume that the distance from the inside face to the neutral axis is $0.5t$.

∴ radius to neutral axis = 5 mm + (0.5 × 2 mm) = 6 mm

Since it is a 90° bend, length BC equals a quarter of the circumference of a circle of radius 6 mm

$$\therefore \text{ length BC } = \frac{2\pi R}{4} = \frac{\pi R}{2} = \frac{6\pi \text{mm}}{2} = 9.4 \text{ mm}$$

∴ blank length = 43 mm + 33 mm + 9.4 mm = 85.4 mm

☐ *Example 16.4*

Determine the blank length of the right-angled bracket shown in Fig. 16.11(b).

Length EF = 80 mm – 1.5 mm – 1.2 mm = 77.3 mm
Length GH = 30 mm – 1.5 mm – 1.2 mm = 27.3 mm

Since the radius in this case is less than twice the material thickness t, we can assume that the distance from the inside face to the neutral axis is $0.33t$.

∴ radius to neutral axis = 1.5 mm + (0.33 × 1.2 mm) = 1.9 mm

$$\therefore \text{ length FG} = \frac{\pi R}{2} = \frac{1.9\pi \text{mm}}{2} = 2.98 \text{ mm, say 3 mm}$$

∴ blank length = 77.3 mm + 27.3 mm + 3 mm = 107.6 mm

Another factor which must be considered in bending operations is the amount of springback. Metal that has been bent retains some of its original elasticity and there is some elastic recovery after the punch has been removed. This is known as springback. In most cases this is overcome by overbending, i.e. bending the metal to a greater extent so that it will spring back to the required angle.

Die-sets

The punch is held in a punch plate and in its simplest form has a step as shown in Fig. 16.12. The step prevents the punch from pulling out of the punch plate during operation.

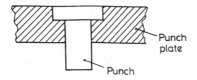

Figure 16.12 Location of punch in punch plate

In order to ensure perfect alignment in the press, the punch plate and the die are secured in a die-set. Standard die-sets are available in steel or cast iron, and a typical example is shown in Fig. 16.13. The top plate carries a spigot which is located and held in the press slide. The bolster contains slots for clamping to the press bedplate. The bolster has two guide pins on which the top plate slides up and down through ball bushes which reduce friction and ensure accurate location between the two parts.

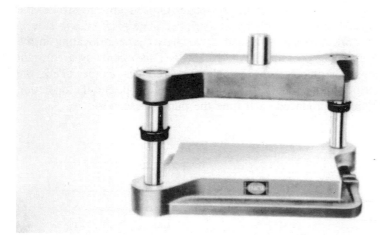

Figure 16.13 Standard die set

The punch plate is fixed to the underside of the top plate and the die is fixed to the bolster, the punch and die being accurately aligned with each other.

The die-set is mounted in the press as a complete self-contained assembly and can be removed and replaced as often as required in the knowledge that accurate alignment of punch and die is always maintained. Change-over times are low, since the complete die-set is removed and replaced by another die-set complete with its punches and dies for a different component.

16.3 Blanking, piercing and bending operations

Simple blanking

As previously stated, blanking is the production of an external shape from sheet metal. In its simplest form, this operation requires one punch and die.

For simple blanking from strip fed by hand, the press tool consists of a die on top of which are attached the strip guides and the stripper plate. The punch is held in the punch plate. The arrangement is shown in section in Fig. 16.14.

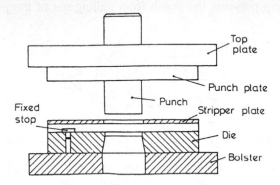

Figure 16.14 Simple blanking tool

To assist in setting up and for subsequent operation, stops are required. For setting up, a sliding stop is pushed in and the strip of work material is pushed against it by hand, Fig. 16.15(a). The punch descends and blanks the first part, the blank falling through the die opening and out through the bedplate. On the upward stroke, the work material is stripped from the punch by the stripper plate. The sliding stop is then retracted and the work material is pushed up to the fixed stop locating in the hole produced in the blanking operation, Fig. 16.15(b). This maintains a constant pitch between blankings. The punch then descends and produces another blank; the punch is raised; the strip is moved forward against the fixed stop; and the operation is repeated. Thus a blank is produced each time the punch descends.

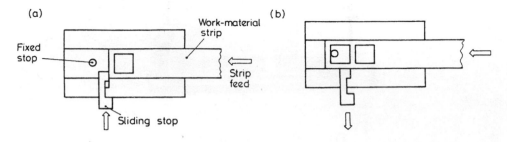

Figure 16.15 Use of stops in simple blanking

Blanking and piercing

Where the required workpiece is to have both an external and an internal shape – for example, a washer – the two operations, i.e. blanking and piercing, can be done by the same press tool. This type of press tool is known as a follow-on tool.

The principle is the same as for simple blanking but the punch plate has two punches fitted – one for blanking and one for piercing – and there are two holes of the required shape in the die, Fig. 16.16. The work material is again in the form of strip, fed by hand.

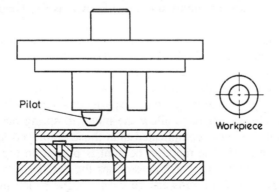

Pilot

Workpiece

Figure 16.16 Follow-on tool

For setting up, two sliding stops are required. The first sliding stop is pushed in and the strip of work material is pushed against it, Fig. 16.17(a). The punches descend and the hole for the first workpiece is pierced. The punches are then raised. The first sliding stop is then retracted, the second sliding stop is pushed in, and the work material is pushed against it, Fig. 16.17(b) – again to maintain a constant pitch. At this stage the pierced hole is now positioned under the blanking punch. The punches once more descend, the blanking punch producing a completed workpiece and at the same time the other punch piercing a hole. The second sliding stop is withdrawn and the work material is now moved forward against the fixed stop, Fig. 16.17(c), again positioning the pierced hole under the blanking punch. The operation is then repeated, a completed workpiece being produced at each stroke of the press.

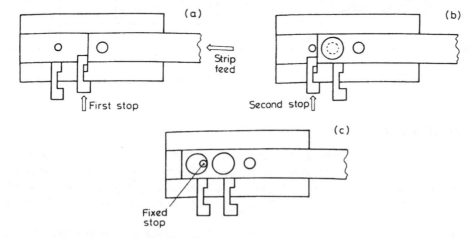

(a)

Strip feed

First stop

(b)

Second stop

(c)

Fixed stop

Figure 16.17 Use of stops in piercing and blanking

Greater accuracy of the inner and outer profiles of a workpiece can be obtained by fitting a pilot in the blanking punch (Fig. 16.16). In this case the fixed stop is used for approximate positioning and is arranged so that the work material is drawn slightly away from it as the pilot engages in the pierced hole.

Bending

Two bending methods are commonly used, one known as vee bending and the other as side bending.

Vee-bending tools consist of a die in the shape of a vee block and a wedge-shaped punch, Fig. 16.18. The metal to be bent is placed on top of the die – suitably located to ensure that the bend is in the correct position – and the punch is forced into the die. To allow for springback, the punch is made at an angle less than that required of the finished article. This is determined from experience – e.g. for low-carbon steel an angle of 88° is usually sufficient to allow the metal to spring back to 90°.

Side-bending tools are more complicated than those employed in vee bending but give a more accurate bend. The metal to be bent is placed on top of the die and is pushed against the guide block, which determines the length of the bent leg. Where the leg length is short, location pins can be used. As the punch plate descends, the pressure pad contacts the surface of the metal in advance of the punch and holds it against the die while the punch forms the bend. The guide plate prevents the punch from moving away from the work material during bending and helps the punch to iron the material against the side of the die, so preventing springback. The arrangement of this type of tool is shown in Fig. 16.19.

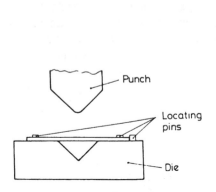

Figure 16.18 Vee-bending tool

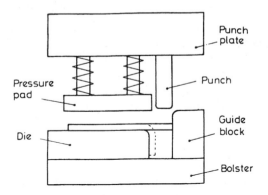

Figure 16.19 Side-bending tool

16.4 Blanking layouts

When parts are to be blanked from strip material, it is essential that the blank is arranged within the strip to gain the greatest economical use of the material by minimising the amount of scrap produced. The final layout will determine the width of strip, which in turn determines the general design and dimensions of the press tool.

The layout may be influenced by subsequent operations such as bending. In this case it is necessary to consider the direction of grain flow, as previously outlined.

It is also necessary to consider the minimum distance between blanks and between blanks and the edge of the strip – this distance must be large enough to support the strip during blanking. Insufficient distance results in a weakened strip which is subject to distortion or breakage, leading to misfeeding. The actual distance depends on a number of variables, but for our purpose a distance equal to the work-material thickness is acceptable.

The material utilisation can be calculated from the area of the part divided by the area of strip used in producing it, given as a percentage. The area of strip used equals the strip width multiplied by the feed distance. An economical layout should give at least a 75% material utilisation.

By virtue of their shape, some parts are simple to lay out whereas others are not so obvious.

Consider a 30 mm × 20 mm blank to be produced from 14 swg. (2.0 mm) material. This would simply be laid out in a straight line as shown in Fig. 16.20.

Since the work-material thickness is 2 mm, the distance between blanks and between blanks and the edges will be assumed to be 2 mm; therefore

strip width = 30 mm + 2 mm + 2 mm = 34 mm

and the distance the strip must feed at each stroke of the press in order to produce one blank is

20 mm + 2 mm = 22 mm

$$\text{Material utilisation} = \frac{\text{area of part}}{\text{area of strip used}} \times 100\%$$

$$= \frac{\text{area of part}}{\text{strip width} \times \text{feed distance}} \times 100\%$$

$$= \frac{30 \text{ mm} \times 20 \text{ mm}}{34 \text{ mm} \times 22 \text{ mm}} \times 100\% = 80\%$$

Now consider the blank shown in Fig. 16.21, to be produced from 19 SWG (1.0 mm) material. The simple layout would be as shown in Fig. 16.22(a). This gives a strip width of 32 mm and a feed of 31 mm, allowing 1 mm between blanks and between blanks and the edges of the strip. Thus the area of strip used per blank would be 32 mm × 31 mm = 992 mm².

Area of blank = 675 mm²

∴ material utilisation $= \dfrac{675 \text{ mm}^2}{992 \text{ mm}^2} \times 100\% = 68\%$

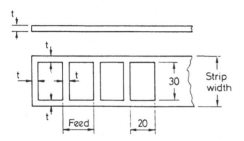

Figure 16.20 Blank layout

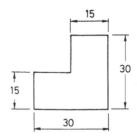

Figure 16.21 Blank

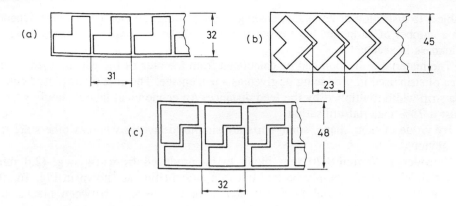

Figure 16.22 Blank layout

An alternative layout with the blanks turned through 45° is shown in Fig. 16.22(b). In this case the strip width is 45 mm and the feed 23 mm. Thus the area of strip used per blank would be 45 mm × 23 mm = 1035 mm².

Area of blank = 675 mm²

∴ material utilisation = $\dfrac{675 \text{ mm}^2}{1035 \text{ mm}^2} \times 100\% = 65\%$

A further alternative layout is shown in Fig. 16.22(c), with the blanks in an alternating pattern. This gives a strip width of 48 mm and a feed of 32 mm, which in this case produces two blanks. Therefore the feed per blank is half this amount, i.e. 16 mm. Thus the area of strip used per blank would be 48 mm × 16 mm = 768 mm².

Area of blank = 675 mm²

∴ material utilisation = $\dfrac{675 \text{ mm}^2}{768 \text{ mm}^2} \times 100\% = 88\%$

This gives the most economical use of the material and would therefore be the obvious choice. However, with this layout the strip would have to be worked in two passes. On the first pass the bottom row would be blanked; the strip would then be turned round and passed through again for the other row to be blanked.

17 Investment casting and shell moulding

Investment casting and shell moulding are employed where greater accuracy is required than can usually be achieved by sand casting, especially with small intricate castings and those made from materials otherwise difficult to work with. The moulds in each process are expendable, as in sand casting, and, in comparison with sand casting, these processes are often referred to as 'precision casting'.

Equipment and materials are more expensive than those required in sand casting and therefore require the production of larger quantities to be economic. The higher costs for small quantities can, however, often be offset by savings as a result of the ability to produce a high degree of dimensional accuracy and surface finish which can reduce or eliminate subsequent machining operations.

Any metal capable of being cast can be cast by these methods, although advantage is most readily gained when using those difficult to work by other methods. Metals for investment casting are the subject of British Standard BS 3146.

17.1 Investment casting

The investment-casting process – also known as the lost-wax process – is one of the oldest casting processes, having been practised for thousands of years. Today highly sophisticated plant, equipment, and materials are employed to produce a large variety of components in materials ranging from high-temperature nickel- or cobalt-based alloys – known as 'super-alloys' – to the non-ferrous aluminium and copper alloys.

The process starts with the production of an expendable wax pattern of the shape required which is then coated with a refractory material to produce the mould, which is allowed to dry. The mould is heated, melting the wax, which is allowed to run out and so produce a cavity. Further heat is applied to fire the mould before pouring the casting metal to fill the cavity left by the melted wax. When the molten metal has solidified, the refractory shell is broken away to release the casting. The casting is cut away from any runners and is dressed or fettled and finished as required.

The expendable wax pattern, exactly the shape and size of the required casting, with allowances for pattern contraction and contraction of the casting during solidification, is produced in a split pattern die, Fig. 17.1. The wax is injected into the pattern die in a plastic state, under pressure, and solidifies quickly. It is then ejected from the pattern die, Fig. 17.2. Ceramic cores can be introduced into the wax pattern for the production of castings which require hollow or complex interior forms. When the wax patterns are small, a number of them are joined by heat welding to a wax runner-and-riser system to produce a pattern assembly for convenience of casting, Fig. 17.3.

Pattern dies are made from a variety of materials, depending on the number of patterns required, their complexity, and the required dimensional accuracy. The die materials used include cast low-melting-point alloys, epoxy resin, and aluminium alloys.

Figure 17.1 Split aluminium die and wax pattern

Figure 17.2 Wax-pattern removal

The completed wax-pattern assembly is first given a primary coat or investment by dipping it in a ceramic slurry of very fine particle size to make a complete one-piece mould, Fig. 17.4. This first coat is extremely important, as it determines the surface finish and final quality of the casting. When the primary coat is dry, the secondary investment or back-up coats are applied to build up the shell to the required thickness. This is done by successive dipping in slurry followed by coating with dry granular refractory or stucco of large particle size – a process known as 'stuccoing'. A completed shell is built up of six to nine such back-up coats, dried between each coat, to give a final thickness of about 6 mm to 12 mm.

After the shell has been formed round the wax pattern and dried, it is dewaxed by heating to about 150°C and allowing the wax to run out, leaving the required shape of cavity inside the mould, Fig. 17.5. The mould is then fired at about 1000°C to attain its full mechanical strength, and the casting metal is poured as soon as the mould is removed from the firing furnace.

Casting may be carried out in air, in an inert atmosphere, or in a vacuum. Vacuum casting is preferred for casting nickel- and cobalt-base 'super-alloys', not only to

Figure 17.3 Wax patterns attached to a runner system

Figure 17.4 Dipping completed wax-pattern assembly

avoid contamination of the alloy but also to ensure perfect filling of the mould and to reduce oxidation of the component after casting.

When the cast metal has solidified, the mould is broken up to release the casting, Fig. 17.6, which are cut away from their runners. The castings are then finished by fettling, finishing using a moving abrasive belt, or vapour blasting where loose abrasive is forced on to the surface under pressure. Any cores present in the casting are removed by using a caustic solution.

A typical range of investment castings is shown in Fig. 17.7.

The dimensional accuracy of investment casting will vary depending upon pattern contraction, mould expansion and contraction, and contraction of the cast metal. In

Figure 17.5 Dewaxing mould shells

Figure 17.6 Refractory removed from cast metal

general, a tolerance of ± 0.13 mm per 25 mm can be achieved on relatively small castings. A surface finish comparable with that of machining can be achieved by the transfer of the excellent finish of the pattern to the surface of the mould cavity by the investment technique.

The majority of investment castings are from a few grams to 10 kg, although castings up to 150 kg have been produced. A wall-section thickness of 1.5 mm can be achieved on production runs, although thinner wall sections are possible.

Although the process can be used for prototype castings, quantities usually produced are within a range from 50 to 50 000.

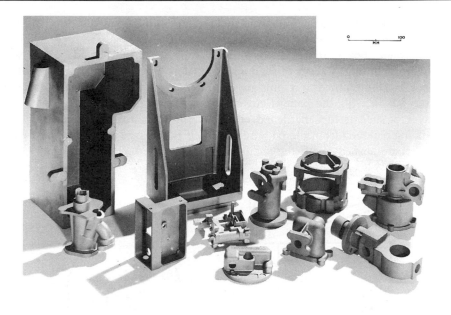

Figure 17.7 Typical investment castings

Advantages of investment casting

- High degree of dimensional accuracy and surface finish.
- Accurate reproduction of fine detail.
- Virtually unlimited freedom of design enabling intricate shapes, both external and internal, to be produced. Sub-assemblies can often be replaced by a single casting.
- Metals can be cast which are difficult or even impossible to work by other methods.
- Reduction or elimination of machining operations, especially on metals which are difficult to machine, resulting in savings in material and machining costs.
- Comparatively small quantities can be produced, as well as large quantities.

17.2 Metals for investment casting

Investment castings can be made from an extremely wide range of ferrous and non-ferrous alloys.

The ferrous range is specified in two parts by British Standard BS 3146, 'Investment castings in metal'. In the non-ferrous range, most of the copper-based alloys specified in BS 1400 (including brasses, bronzes and gun metals) and a number of the aluminium alloys specified in BS 1490 can be satisfactorily cast.

17.3 Shell moulding

The shell-moulding process uses a thin expendable mould or 'shell' made from a fine silica sand bonded with a thermosetting resin. The fineness of the sand influences the surface of the finished casting.

The shell is produced by heating a metal pattern of the shape to be cast to a temperature of about 230°C to 260°C and covering it with the sand-and-resin mixture. This is done by locating the heated pattern on the open end of a dump box, mounted on trunnions, which contains the sand-and-resin mixture, Fig. 17.8(a). When the dump box is inverted, the mixture falls on to the heated pattern which melts the resin through a layer of sand approximately 10 mm thick, Fig. 17.8(b). The dump box is then returned to its upright position and surplus mixture falls back into the dump box, leaving the pattern covered with a layer of sand bonded with resin. This layer or shell is then hardened or 'cured' by the further application of heat, before being removed from the pattern. The shell is removed with the aid of ejector pins which are pushed to lift the shell clear of the pattern, Fig. 17.8(c).

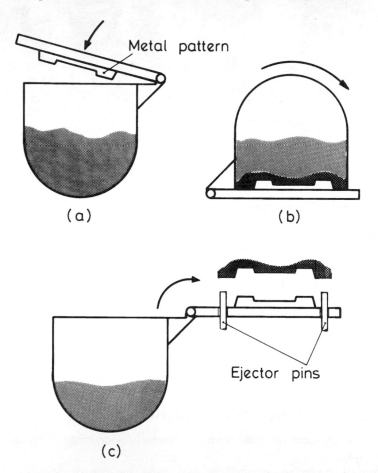

Figure 17.8 Stages in producing a shell mould

Patterns suitable for shell moulding can be made from any material capable of withstanding oven temperatures of about 400°C and of transferring heat to the moulding material. The materials most widely used are cast iron and steel. These metal patterns are more expensive than the wooden patterns used with green-sand casting and so, to be economic, the process requires larger numbers of castings – seldom less than 200.

When the shells required to form a complete mould have been cured, they are placed together and secured ready for metal pouring. They can then be used immediately or stored for later use.

Before pouring, large shells may have to be supported to avoid distortion under the heavy mass of molten metal.

Molten metal is poured into the prepared shell mould to form the casting, the surface of which will be free of blemishes as air and gases can readily escape due to the permeability of the shell. When the cast metal has solidified, the shell is broken away to reveal the casting. Figure 17.9 shows the metal pattern, shell mould, cores, and resulting casting.

Figure 17.9 Shell-moulding pattern, mould, cores and casting

Shell moulding can generally be expected to produce small castings more accurately and with smoother surfaces than other forms of sand casting. It is comparable with gravity die-casting in aluminium but not as good as pressure die-casting or investment casting.

An accuracy within 0.25 mm can be maintained on dimensions up to 100 mm in the same mould half. Accuracies of about 0.4 mm could be expected for similar dimensions across the mould parting line. These accuracies will often eliminate machining operations, with consequent cost savings.

The smooth surface finish provides a good base for paint finishes or an improved appearance if left in the as-cast condition.

Shell moulding allows intricate shapes to be produced with minimum taper or draft angle. An angle of $\frac{1}{2}°$ to $1°$ is usually sufficient, compared with the $2°$ to $3°$ needed with green-sand casting.

Although the sand used is more expensive than green sand, the volume used is less, due to the thin shell, and this can show savings in the amount purchased, stored, and handled.

It has been estimated that, on average, a relatively simple shell-moulded casting will cost around 10% to 15% more than the same casting produced in green sand. However, by taking advantage of the greater accuracy and reducing or eliminating machining operations, the use of a shell-moulded casting can often result in a cheaper finished component.

The shell-moulding process can be used to produce castings in any metal capable of being cast. Materials commonly used include bronze, aluminium, magnesium, copper, and ferrous alloys. Although the patterns and sand are relatively expensive, the use of this process can usually be justified, in comparison with green-sand casting, for complex castings where high accuracy and definition are required on reasonable-size repeat batches.

18 Die-casting

Die-casting is the name given to the production of castings by processes which make use of permanent metal moulds or 'dies'. The term 'permanent' is used since the moulds can be used to produce thousands of castings before they require replacement, unlike those used in sand casting, where each mould is destroyed to remove the casting.

The basic principle of all die-casting is for the die, which contains a cavity of the required shape, to be filled with molten metal. The die is usually in two parts, but may often have movable pieces of 'cores', depending on the complexity of shape being produced.

With the two die halves securely locked in the closed position, molten metal is introduced into the die cavity. When the molten metal has solidified, the die halves are opened and the casting is removed or ejected. The die halves are then closed and the operation is repeated to produce the next die-casting. With fully automatic machines this complete cycle can be done extremely quickly.

Each fill of metal is known as a 'shot'. It is the way in which filling is done which distinguishes the different die-casting methods.

18.1 Gravity die-casting

Gravity die-casting is the simplest and most versatile of the die-casting methods. The molten metal is poured into the die, using the force of gravity to ensure that the die cavity is completely filled in the same manner as for sand casting. The die contains the necessary cavity and cores to produce the required shape of casting, together with the runners, gates, risers, and vents needed to feed the metal and allow air and gases to escape, Fig. 18.1. Apart from the dies, very little additional equipment is required, the dies being arranged near to the furnaces where the metal is melted and from where it can be transferred by ladle.

This process provides castings of greater accuracy than sand casting, due to the use of metal moulds which can be used to produce thousands of castings before they require replacement. The most commonly used material for the manufacture of dies for gravity die-casting is close-grained cast iron. Any cores required are usually made of heat-resisting steel. The dies are more expensive than sand-casting moulds but cheaper than those required with high-pressure die-casting methods.

To allow for runners and risers, the volume of material required can amount to over twice the volume of the actual component. Although the surplus can be recovered, it is still wasteful in its removal, dressing the component (or fettling using a hand-held grinder), and the energy required in remelting.

The output from gravity die-casting varies with casting size and may be as high as 25 shots/h, but the process requires greater manpower in operation and filling than other die-casting methods. It is generally used for lower-quantity production with aluminium and magnesium alloys and brass, for castings up to about 23 kg.

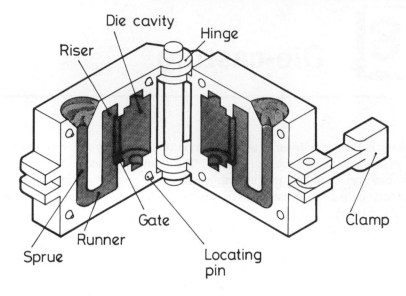

Die cavity

Hinge

Riser

Gate

Clamp

Runner

Locating
pin

Sprue

Figure 18.1 Gravity die-casting

18.2 Low-pressure die-casting

The low-pressure die-casting process involves holding the molten metal to be cast in a crucible around which are arranged the heating elements necessary to keep the metal molten. The crucible and elements are contained inside a sealed pressure vessel in the top cover of which is a riser tube whose lower end is immersed in the molten metal. Low-pressure air (0.75 bar to 1.0 bar) is introduced into the crucible above the surface of the molten metal and forces the metal up the riser tube and out through the top end. This top end is attached to the bottom or fixed half of the die, which is designed to allow air to escape as the molten metal enters. When the metal has solidified in the die, which is air-cooled, the air pressure is released and the remaining molten metal in the riser tube drops.

The casting is removed by raising (usually by means of a hydraulic cylinder) the moving platen to which the top or moving half of the die is attached. The dies are designed so that the moving half contains the male shape, and so, due to shrinkage of the metal, the casting clings to this half. When the moving platen nears the end of its stroke, striker pins on the ejector mechanism contact the underside of the fixed top platen, pushing the ejector pins which release the casting from the die. In some cases a stripper ring is used to release the casting, but the principle is the same.

The moving platen is then reversed to close the two halves of the die ready for the next casting to be produced.

A diagrammatic view of the machine set up with the dies in the closed position is shown in Fig. 18.2.

The dies are cheaper than for high-pressure methods since they can usually be made of cast iron, but they are sometimes more expensive than for gravity die-casting. The capital outlay for machines is higher than for the gravity but lower than for the high-pressure method.

Due to the need for only a single feed, less metal is used and therefore less dressing (or fettling) is required. This results in minimal distortion and makes the process suitable for the production of large flat castings. Components produced by

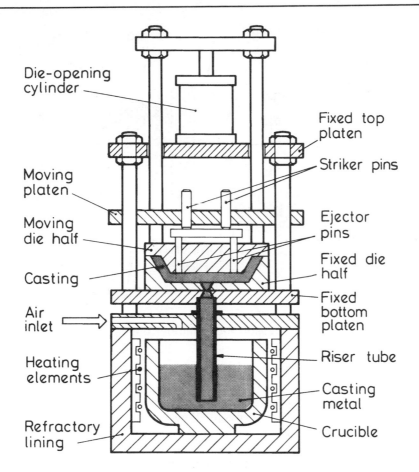

Figure 18.2 Low-pressure die-casting machine

this method can be both lighter and thinner than with gravity die-casting, as the molten metal is introduced under pressure rather than relying solely on gravity to fill the die.

While low-pressure die-casting is not necessarily quicker than gravity die-casting, it does give a more dense-structured casting and lends itself to automation, where one man can operate more than one machine with a consequent higher output per man. This technique is used for medium-volume production, most commonly in aluminium alloys, for castings up to about 10 to 12 kg at rates of around 12 shots/h, depending on casting size.

Castings in excess of 60 kg can be produced by this method.

18.3 High-pressure die-casting

Two principal types of machine are used in the production of die-castings by high-pressure methods: hot-chamber and cold-chamber machines.

Hot-chamber machine

A section through a machine of this type is shown in Fig. 18.3. As shown, the furnace and the crucible containing the molten metal are contained within the machine.

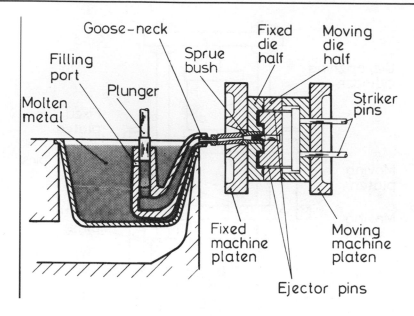

Figure 18.3 High-pressure hot-chamber die-casting machine

The casting cycle starts with the moving machine platen, to which one die half is attached, being pushed forward, usually by a pneumatic cylinder, until it locks against the other die half attached to the fixed machine platen. Only when the dies are completely locked does the safety system allow the plunger to be depressed.

As the plunger is depressed, it covers up the filling port and so prevents molten metal escaping back into the crucible. The molten metal is forced through the goose-neck into the sprue bush and fills the die cavity. To enable air to escape from the die cavity, small channels known as 'vents' are machined from the cavity across the die face. These vents must be deep enough to allow the air to escape but shallow enough to prevent metal escaping from the die. Usually a depth of 0.13 mm to 0.2 mm is sufficient.

When the die is filled, the plunger is returned to the 'up' position, the molten metal in the nozzle flows back into the goose-neck, and the injection cylinder refills through the filling port.

Meanwhile the molten metal in the die has solidified, due to the lower temperature of the die. The die is maintained at the correct temperature by circulating cold water through it.

When the plunger reaches the 'up' position, the mechanism to unlock the dies is operated and the moving machine platen and the die half attached to it are withdrawn. The dies are constructed so that the male form and cores are on the moving die half, and the casting shrinks on to this half and comes away with it.

As the moving machine platen continues to open, striker pins bearing on the ejector plate strike an ejector block on the machine which pushes the ejector pins forward, releasing the casting. The casting falls through an opening in the machine base, down a chute, and into a container or quenching tank. The dies are then closed and the cycle is repeated.

This cycle can take place as rapidly as 500 times per hour, or more on high-production machines.

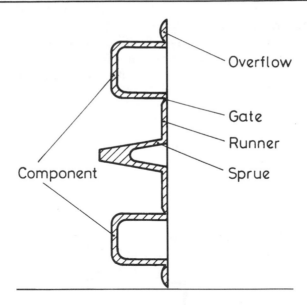

Figure 18.4 Spray from high-pressure hot-chamber die-casting machine

The casting as ejected from the machine is known as the 'spray' and contains the component, sprue, runner, and overflow as shown in Fig. 18.4.

The 'runner' is a channel in the die through which the molten metal fills the die cavity. The term is also applied to the metal that solidifies in this channel. The 'gate' is that part of the die through which the metal enters the die cavity from the runner.

The 'sprue' is the metal attached to the runner that solidifies in the sprue bush. The 'overflow' is a recess with entry from the die cavity, to assist in the production of sound castings and ensure complete filling of the cavity.

Metal in the region of the gate and to the overflow is made thin so that the runner and overflow can be easily and quickly broken off the casting. These are placed in such a position as not to impair the finished appearance of the casting.

Hot-chamber machines are most widely used with zinc alloy at pressures in the region of 100 bar. The metal temperature in the crucible is between 400°C and 425°C, and the die temperature, which should be the lowest that will give castings of good quality, is usually from 180°C to 260°C.

The cost of dies, which are made from heat-resisting steel, is high and makes this process uneconomical for small numbers of components. Annual requirements of 20 000 or more are generally necessary to make production economical, but this would depend on the complexity of the component and whether more than a single cavity could be incorporated in the die. Dies with more than one cavity are known as multi-impression dies.

This process is highly automated to produce rapid cycle times. A very high surface finish can be obtained on dimensionally accurate castings requiring little or no machining. Thin sections can be produced down to 1 mm or even 0.5 mm, which result in lighter castings and a reduction in the amount of metal required.

Cold-chamber machine

A section through a machine of this type is shown in Fig. 18.5. Unlike in the hot-chamber type, the metal is melted and held in a furnace away from the machine.

Again, one die half is attached to the fixed machine platen and the other to the moving machine platen.

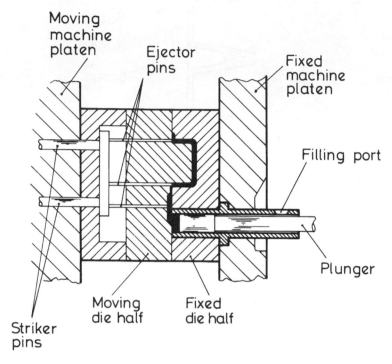

Figure 18.5 High-pressure cold-chamber die-casting machine

With the dies locked in the closed position, a measured quantity of molten metal is fed from the holding furnace into the injection cylinder. This operation can be automated but is usually done by hand, using a ladle of a size appropriate to the shot size required for the casting. The plunger is pushed forward, usually by means of a hydraulic cylinder, forcing the molten metal into the die cavity, where it quickly solidifies. Pressure on the metal varies between 350 bar and 3500 bar. The dies are opened and the casting is released by the ejector mechanism in the same way as in the hot-chamber machine.

The plunger is then withdrawn and the dies are closed ready for the next cycle. The cycle time is not as fast as for the hot-chamber machine.

In this machine there is no sprue, since the die is mounted at the end of the cylinder. Instead, the metal not used in filling the cavity remains in the cylinder and is attached to the casting when the spray is removed from the die. This excess metal attached to the runner is known as the 'slug', Fig. 18.6.

Cold-chamber machines are more commonly used with the higher-melting-point aluminium alloys poured at about 650°C to 670°C and with magnesium alloys poured at about 680°C.

Rates of production in the region of 80 to 100 shots/h can be achieved with hand-filled machines, but these can be increased by about 40% by using mechanised metal-feed systems.

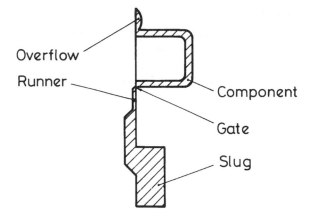

Figure 18.6 Spray from high-pressure cold-chamber die-casting machine

18.4 Die-casting metals

Although techniques for pressure die-casting in steel have been developed, the metals most widely used are certain non-ferrous alloys which have relatively low melting points.

Zinc alloys

The compositions of zinc die-casting alloys are set out in BS 1004. Two alloys are specified: alloy A and alloy B whose composition is similar except that alloy B contains about 1% copper.

In most cases alloy A is used for the production of zinc die-castings and has the advantage over alloy B of greater dimensional stability and rather more resistance to corrosion. Alloy B is sometimes preferred due to its slightly greater strength and hardness, e.g. in zip-fastener slides and gear wheels.

Zinc alloys can be easily cast since they are extremely fluid even at relatively low temperatures, which allows the molten metal to flow into the most intricate shapes and permits thin sections to be cast.

Alloy A solidifies at 382°C and is generally cast at temperatures from 400°C to 425°C. This relatively low temperature enables smaller tolerances to be maintained and leads to longer die life, since the dies are not subjected to repeated high thermal shock.

The low casting temperature of these alloys makes them particularly suitable for use with hot-chamber high-pressure machines which are fully automatic and allow high rates of production.

Zinc alloys have good resistance to corrosion, and die-castings can be left as-cast. However, in situations which are particularly corrosive, chromating or anodising is used for added protection. Where functional and decorative finishes are required, the most widely used is chromium plating, and the best results are obtained when an initial layer of copper is followed by layers of nickel and chromium. Other finishes such as painting and plastics coating can also be applied.

Zinc alloys possess good machining qualities and, although zinc die-casting can be produced to fine tolerances, machining can be easily carried out where necessary.

Aluminium alloys

The major advantages of aluminium alloys are lightness and high electrical and thermal conductivity. Aluminium alloys are about 40% as dense as zinc alloys and, although not generally as strong, many aluminium alloys can be heat-treated to give comparable mechanical properties. Section thicknesses, however, have generally to be thicker to obtain equivalent strengths.

The melting temperature is about 600°C, with casting temperatures about 650°C. Because of the higher melting temperature required, aluminium-alloy die-castings are less close in tolerance than zinc die-castings but can be used at higher working temperatures.

The compositions of aluminium alloys are specified in BS 1490 and the two most widely used in low-pressure and high-pressure cold-chamber machines are LM 2 and LM 24. Where intricate and thin walled castings requiring good corrosion resistance are required, LM 6 is widely used in gravity and pressure die-casting.

Due to the wider tolerances required with aluminium-alloy die-castings, it is usual to machine in order to obtain acceptable limits of size. Machining can readily be carried out using conventional machine tools and cutting tools, but with the higher cutting speeds associated with aluminium and its alloys.

Finishing of aluminium-alloy die-castings is usually for decorative purposes and includes anodising, which can be dyed a wide variety of colours; painting; and electroplating. The more usual electroplating material is chromium, which involves fairly complicated pre-treatment to prevent the formation of an oxide film which would make it difficult for the plated metal to bond to the aluminium.

Magnesium alloys

Magnesium alloys are the lightest of the casting metals, being about 60% as dense as aluminium alloy, and are used where weight saving is important, e.g. for portable saws, cameras, projectors, and automobile components including specialised car wheels. The specification of these alloys is contained in BS 2970.

Because of choking of the nozzle due to the disturbance of the flux cover on the molten metal, this material cannot be used in hot-chamber machines. It is used satisfactorily with cold-chamber machines, where it is poured at about 680°C.

Magnesium alloys machine extremely well, but care must be taken not to allow the cutting edge of the tool to become dull, as the heat generated may be sufficient to ignite the chips.

Magnesium alloys do not retain a high lustre like the aluminium alloys and cannot be electroplated with the same ease as the zinc alloys. Finishing is usually achieved by painting or lacquering. Chromate treatments are used to protect against corrosion, but have little decorative value.

Copper alloys

Of the many copper alloys, only 60/40 brass and its variants with a melting temperature of about 900°C combine acceptable mechanical properties with a casting temperature low enough to give reasonable die life and is contained in BS 1400.

Apart from its high injection temperature of about 950°C, brass is an extremely good die-casting alloy, reproducing fine detail and flowing readily into exceptionally thin sections. By keeping all the sections thin, heat transfer to the die can be very greatly reduced and die life be extended.

Lead and tin alloys

These low-melting-point alloys were the first alloys to be die-cast, originally for printer's type. Lead alloyed with antimony, sometimes with small additions of tin, has a melting point of about 315°C and can be cast to very close tolerances and in intricate shapes. The castings have low mechanical properties and are used mainly for their density, e.g. car-wheel balance masses, and corrosion-resistance, e.g. battery-lead terminals.

Several tin-based alloys, usually containing lead, antimony, and copper, with a melting point of about 230°C, are also die-cast where the very highest accuracy is required and great strength is not of importance. Their excellent corrosion-resistance to moisture makes them suitable for such components as number wheels in gas and water meters, and they have also found use in small complex components of electrical instruments.

18.5 Special features of die-castings

Of all the methods used in manufacturing, die-casting represents the shortest route from molten metal to finished part. To take greatest advantage of the special features of die-casting – such as accuracy, rate of production, thinness of section, lack of machining, and fine surface finish – great attention must be given to the design of parts intended for die-casting. It is advisable for the designer to consult the die-caster at an early stage of design in order that those features which permit ease of production can be incorporated.

Section thickness

Sections should be as thin as possible consistent with adequate strength. Thin sections reduce metal cost and allow the casting to solidify faster in the die, thus shortening the production cycle. They also result in lighter components. Small zinc die-castings, for instance, can be produced with wall sections down to 0.5 mm, though with larger zinc die-castings a thickness of 1 mm is more general. Additional strength of a thin section can be achieved by providing ribs in the required position.

The sections of a die-casting should always be as uniform as possible – sudden changes of section affect the metal flow and lead to unsound areas of the casting. Differences in the rate of cooling between thick and thin sections produce uneven shrinkage, causing distortion and stress concentrations.

Bosses are sometimes required to accommodate screws, studs, pins, etc. and, if designed with a section thicker than an adjacent thin wall, will also cause unequal shrinkage. This can be minimised by making the variation in thickness as small and as gradual as conditions permit.

Die parting

The die parting is the plane through which the two halves of the die separate to open and close. It is usually across the maximum dimension. The designer should visualise the casting in the die and design a shape which will be easy to remove.

When the two die halves are closed, there is always a small gap at the two faces, into which metal will find its way due to pressure on the metal. This results in a small ragged edge of metal known as a 'flash', which has subsequently to be removed, usually by means of a trimming tool. The position of the flash has to be arranged so that it can be removed efficiently without leaving an unsightly blemish on the finished casting.

Ejector pins

Ejector pins are used in the moving die to release the casting. The ejector pins will leave small marks on the surface of the casting and should be positioned so that these will not appear on a visible face of the finished casting. If a casting has a face which is to be machined, then, where possible, the ejector-pin marks should be arranged on that face, to be removed by the machining.

Draft angle

To allow the casting to release easily from the die, a wall taper or draft angle, normally between 1° and 2° per side, is provided. With shallow ribs more taper is required, about 5° to 10°.

Undercuts

A part which contains a recess or undercut requires slides or moveable cores in the die, otherwise the casting cannot be ejected. These slides and moveable cores greatly increase die costs and slow down the rate of production. As a general rule, the design of a part should avoid undercut sections.

Corners

Sharp internal corners on castings are always a source of weakness, and should be avoided by the use of a blend radius or fillet. For example, it is common practice with high-pressure casting of zinc alloys to have a minimum radius of 1.6 mm on inside edges. A slight radius on the outside corners of castings reduces die cost.

The provision of radii within the die cavity is beneficial to the flow of molten metal and the production of sound castings.

Lettering

When die-cast lettering, numerals, trade marks, diagrams, or instructions are required, they should be designed as raised from the casting surface. This reduces die cost, as it is easier to cut the design into the die surface than to make a raised design on the die surface.

Threads

Threads can be cast on high-pressure machines but should be specified only where their use reduces cost over that for machine-cut threads. Cast internal threads under 20 mm in diameter are rarely economical.

Inserts

It is sometimes desirable to include inserts to obtain features in a casting which cannot be obtained from the cast metal. This may be done to provide

- additional strength,
- locally increased hardness,
- bearing surfaces,
- improved electrical properties,
- passages otherwise difficult to cast,

- passages intended to carry corrosive fluids,
- facilities for soldered connections,
- means of easier assembly.

Usually inserts are cast in place, but there are instances in which they are applied after casting, in holes cast for that purpose. The object of casting the insert in place is either to anchor it securely or to locate it in a position where it could not be placed after casting.

The insert material must be able to withstand the temperature of the molten metal and is usually steel, brass, or bronze.

Inserts which are cast in position have to be manufactured with a small tolerance, otherwise they will not locate accurately in the die or in the casting. In some instances they may be quite expensive to manufacture, and their use tends to slow down the casting process. They should therefore be used only where distinct advantages can be obtained.

Very small inserts are difficult to place in the die and should be avoided. Large inserts can lead to distortion resulting from different coefficients of expansion and should be used with caution.

When inserts are cast in place they are located in the mould cavity, molten metal flows around them and solidifies, and the insert becomes part of the casting. However, there is little or no bond between the casting and the insert other than the mechanical effect of casting shrinkage, so the insert must contain some feature to provide positive anchorage. The simplest methods make use of knurling, holes, grooves, or flats machined on the insert. Some examples are shown in Fig. 18.7.

Figure 18.7(a) shows a location pin anchored by means of a diamond knurl on the outside diameter.

Figure 18.7(b) shows a flat electrical connector anchored by its shape and by molten metal flowing through the hole.

Figure 18.7(c) shows a threaded bush anchored due to its hexagonal shape and the groove round the outside.

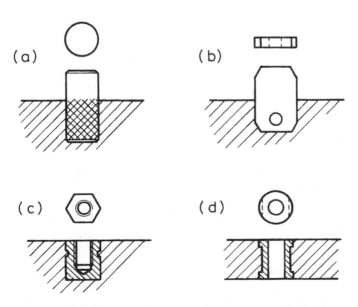

Figure 18.7 Inserts used in die-casting

Figure 18.7(d) shows a plain circular bush anchored due to the two flats machined in the outside diameter.

18.6 Advantages and limitations of die-castings

Advantages

- Die castings can be produced to good dimensional accuracy. Actual values depend on the die-casting method, size of dimension, and material being cast; e.g. a 25 mm dimension pressure-die-cast in zinc alloy can be produced to a tolerance of 0.1 mm if the dimension is in one die half or 0.26 mm if the dimension is across the die parting.
- Due to good dimensional accuracy, machining operations can be reduced or eliminated. Holes, recesses, chamfers, etc. can be cast in position.
- Where machining has to be carried out, the consistent size and shape of die-castings enable accurate location in jigs and fixtures.
- Complex-shaped die-castings can be produced, giving designers a certain freedom to achieve attractive styling, and can be shaped to give localised strength where required.
- The ease with which complex parts can be produced enables die-castings to replace parts which are extremely difficult to produce by machining and to replace a number of assembled parts with a single casting and so save on assembly costs.
- Thin walls can be produced, resulting in
    i) a lighter casting;
   ii) material-cost savings, since less material is used;
  iii) reduced cost per casting, since thin walls allow higher rates of production.
- Smooth surfaces of die-castings from metal moulds reduce or eliminate pre-finishing operations such as buffing and polishing prior to finishing operations such as electroplating and painting.
- The cost of die-castings can be reduced by using multi-impression dies.
- A wide range of finishes can be given to die-castings. Choice depends on the die-cast materials and includes electroplating, vacuum metallising, painting, lacquering, plastics coating, anodising, and chromating.
- Die-castings can be produced with cast-in inserts of other metals. Inserts such as tubes, heater elements, fasteners, and bushes are widely used:
- The cost of die-castings can be low, due to very rapid fully automated production over long periods. High-pressure machines, for example, can operate at rates in excess of 500 shots/h.
- A wide range of sizes of die-castings can be produced, from a few grams to over 60 kg depending on the die-casting method.

Limitations

- Large quantities of castings are required to offset the high cost of dies and casting machines.
- Die-castings can be produced only from a limited range of the lower-melting-point non-ferrous alloys.
- There are limitations on the maximum size of die-casting, due to metal temperatures and pressures.

18.7 Choice of a die-casting process

There are a number of factors which influence the choice of a die-casting method. These are set out in Table 18.1 opposite each of the die-casting methods.

For example, if the component under consideration is to be made from aluminium alloy and only 2000 per year are required, the choice would be limited to gravity die-casting. If, however, the quantity required was 50 000 per year then the high-pressure cold-chamber method would be appropriate.

Use of Table 18.1 is intended only as a guide, as it is difficult to give precise values. For instance, casting masses depend on the material being cast, accuracy depends on size, and so on.

Table 18.1 Factors influencing choice of die-casting process

Die-casting method	Economic run per year	Accuracy	Casting mass	General minimum thickness of section	Type of metal cast	Rate of prodn	Surface finish	Die costs	Capital cost of equipment
Gravity	1000	Good	Up to about 23 kg or more	6 mm	Al alloys Cu alloys Mg alloys	Medium	Good	High	Nil to very little
Low-pressure	5000	Good	Up to about 12 kg or more	5 mm	Al alloys	Low	Very good	Usually greater than for gravity	High
High-pressure Hot	20 000	Excellent	Less than 1 g to about 5 kg or more	1 mm	Zn alloys	Very high	Excellent	Usually 2 to 3 times greater than for low pressure	Very high
High-pressure Cold				2 mm	Al alloys Cu alloys Mg alloys	High			

Appendices: screw-thread forms

Appendix 1: basic form for ISO metric threads, Fig. A1

Major diameter in millimetres (first-choice sizes)	Pitch in millimetres	
	Coarse-pitch series	Fine-pitch series
1.0	0.25	0.2
1.2	0.25	0.2
1.6	0.35	0.2
2.0	0.4	0.25
2.5	0.45	0.35
3.0	0.5	0.35
4.0	0.7	0.5
5.0	0.8	0.5
6.0	1.0	0.75
8.0	1.25	0.75
8.0	–	1.0
10.0	1.5	0.75
10.0	–	1.0
10.0	–	1.25
12.0	1.75	1.0
12.0	–	1.25
12.0	–	1.5
16.0	2.0	1.0
16.0	–	1.5
20.0	2.5	1.0
20.0	–	1.5
20.0	–	2.0
24.0	3.0	1.0
24.0	–	1.5
24.0	–	2.0

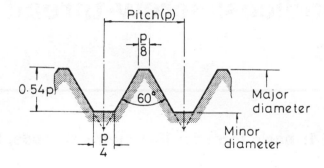

Figure A1 Basic form for ISO metric threads

Appendix 2: basic form for Unified threads (UNC and UNF), Fig. A2

Major diameter in inches	Threads per inch		Pitch in inches	
	UNC	UNF	UNC	UNF
1/4	20	28	0.05	0.036
5/16	18	24	0.055	0.042
3/8	16	24	0.062	0.042
7/16	14	20	0.071	0.05
1/2	13	20	0.077	0.05
9/16	12	18	0.083	0.055
5/8	11	18	0.091	0.055
3/4	10	16	0.100	0.062
7/8	9	14	0.111	0.071
1"	8	12	0.125	0.083

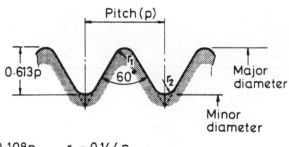

$r_1 = 0.108p$ $r_2 = 0.144p$

Figure A2 Basic form for Unified threads (UNC and UNF)

Appendix 3: basic form for Whitworth threads (BSW, BSF, and BSPF), Fig. A3

Major diameter in inches	Number of threads per inch (t.p.i.)			Pitch in inches			Outside diameter in inches
	BSW	BSF	BSPF	BSW	BSF	BSPF	BSPF
1/8	40	–	28	0.025	–	0.036	0.383
3/16	24	32	–	0.042	0.031	–	–
7/32	–	28	–	–	0.036	–	–
1/4	20	26	19	0.050	0.038	0.053	0.518
9/32	–	26	–	–	0.038	–	–
5/16	18	22	–	0.055	0.045	–	–
3/8	16	20	19	0.062	0.050	0.053	0.656
7/16	14	18	–	0.071	0.055	–	–
1/2	12	16	14	0.083	0.062	0.071	0.825
9/16	12	16	–	0.083	0.062	–	–
5/8	11	14	14	0.091	0.071	0.071	0.902
11/16	11	14	–	0.091	0.071	–	–
3/4	10	12	14	0.100	0.083	0.071	1.041
7/8	9	11	14	0.111	0.091	0.071	1.189
1"	8	10	11	0.125	0.100	0.091	1.309

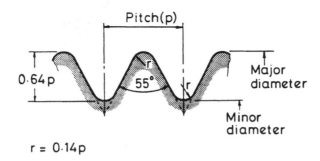

Figure A3 Basic form for Whitworth threads (BSW, BSF and BSPF)

Appendix 4: basic form for British Association threads (BA), Fig. A4

BA designation number	Major diameter in millimetres	Pitch in millimetres
0	6.0	1.00
1	5.3	0.90
2	4.7	0.81
3	4.1	0.73
4	3.6	0.66
5	3.2	0.59
6	2.8	0.53
7	2.5	0.48
8	2.2	0.43
9	1.9	0.39
10	1.7	0.35

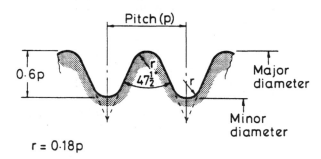

Figure A4 Basic form for British Association threads (BA)

Index

abrasion resistance 85
abrasive wheels 145, 147
abrasives 148
ABS 215
adhesives 183
aluminium 202
  alloys 274
aminos 214
angle plate 47
annealing 196
anodising 208
apron 125
arbor support 157
arc welding 182
arcing 16

BA threads 173
base 154
bed 121
bending 61, 251
  tool 64
blanking 248
  layouts 256
blind rivets 176
bolts 171
bond 149
boring 134
brazing 179
  alloys 179
brittleness 192
bronze 201
BSF threads 173
BSP threads 173
BSW threads 173

cadmium plating 207
cast iron 198–200
cemented carbide 87
centre
  drill 133
  head 50
  punch 53
centres 130
ceramics 88
chisels 34

chromating 209
chromium plating 208
chuck
  4 jaw 127
  milling 162
  collet 127
  3 jaw 126
clamping 105
clamps 53
clearance 89
co-ordinates 44
column 154
combination set 48
compound slide 124, 135
compression moulding 225
connections 187
cope 233
copper 200
  alloys 274
cores 235
corrosion 206
counterbore 107
countersink 107
critical temperature 195
cross slide 124
cubic boron nitride 88
cutter, mounting 161
cutters
  arbor-mounted 158
  screwed-shank 160
cutting
  fluids 97
  sheet-metal 61
  speed 95
cutting-tool materials 85

datum 44
density 190
depth micrometer 79
development (sheet-metal) 65
dial
  caliper 71
  indicators 80
diamond 89

die
  casting 267–9
  casting alloys 203
  parting 275
  sets 253
  sheet-metal 63
  threading 40
dividers 51
double insulation 16
drag 233
drawing 239
drift 104
drill
  chuck 103
  grinding 94
drilling
  machine 102
  plastics 111
  sheet metal 111
drills 92, 106
ductility 192, 233

earthing 16
ejector pins 276
elasticity 193
electric
  burn 15
  shock 15
  connections 187
  resistivity 191
electronic caliper 71
electroplating 207
encapsulation 221
engineer's square 48
explosion 16
external micrometer 76
extrusion 238
eye protection 10

faceplate 128
files 30–2
filing 32
fire 15, 21
  extinguishers 20, 24
  fighting 23
  precautions 22
  prevention 22
fluidity 233
fluxes 177, 180
fly press 63, 243
folding machine 64
foot protection 11
forging 240
form tool 135

galvanising 208
gas
  containers 20
  welding 183
gold 204
grade 148
grain size 148
graphite 206
gravity die-casting 267
grinding wheels 145, 147
GRP 214
guillotine 63

hacksaw 33
hammers 37
hand protection 11
hardening 196
hardness 192
head protection 11
headstock 121
Health and safety
  commission 1
  executive 1
heat
  capacity 189
  treatment 195
hermaphrodite calipers 51
high-pressure die-casting 269
high-speed steel 86
HSW Act 1

injection moulding 227
inserts 228, 276
internal micrometer 79
investment casting 259

jacks 47
joint design 178, 181

knee 155

ladders 28
lead 203
  alloy 275
lettering 276
lifting loads 11
linear expansion 189
lost wax 259
low-pressure die-casting 268

machine screws 169
machinery 28
magnesium alloys 274
MAGS welding 183

malleability 193, 233
mandrel 132
marking dye 50
material utilisation 257
mechanical connections 187
melting point 190
metric threads 172
micrometers 76
milling
    chuck 162
    cutters 158
MMA welding 182
morse taper 104

neat cutting oils 98
neutral axis 251
nickel plating 208
noise levels 15
normalising 196
nuts 171
nylon 206, 216

overarm 157

painting 209–12
parallels 46
pattern 235
perspex 216
phenolics 214
phosphating 209
phospher bronze 205
piercing 248
pins (spring tension) 172
plasticity 233
plastics
    heat bending 219
    machining 218
    welding 217
platinum 204
polypropylene 216
polystyrene 215
polythene 215
power press 244
powered hand tools 40–3
press-tool design 248
protective clothing 11
protractor head 48
PTFE 205, 216
punch 63
PVC 215

rake 89
reamer 107
reaming 134
red hardness 85

resinoid bond 149
rivets 175
rolling 236
rotary table 163
rubber bond 150

saddle 123, 156
safety
    committees 4
    representatives 3
sand 235
    casting 233
scrapers 36
screw cutting 137
screwdrivers 38
screws 169, 170
scriber 51
self-tapping screws 170
semi-synthetic fluids 99
shaping tools 91
shears 61
shell moulding 263
silver 204
snips 61
socket screws 170
soft soldering 177
soldered connections 187
solders 178
solid rivets 175
soluble oils 98
spindle 156
spotface 108
square head 50
steadies 131
steel
    plain carbon 193
    rule 53
stellite 86
strength 192
stripping 250
structure 149
stub arbor 162
surface
    gauge 51
    plate 46
    table 46
synthetic fluids 99

table 156
TAGS welding 183
tailstock 123
tang 104
taper turning 134, 136
taps 38, 109
tempering 197

thermal conductivity 191
thermoplastics 215
thermosetting plastics 214
threads 172–3
tin
 alloys 275
 plating 208
tool
 holding 103
 setting 133
toolmakers buttons 129
top slide 124, 135
toughness 85, 193, 233
trammels 51
transfer moulding 226
trepanning tool 108
tubular rivets 175
tungsten 204
turning
 operations 132
 tools 91
twist drills 92, 106

undercuts 276
unified threads 173

vee blocks 47
vehicle connections 187
vernier
 bevel protractor 73
 caliper 70
 depth gauge 73
 height gauge 51, 72
 protractor scale 74
 scale 69
vice 163
vitrified bond 149

washers 172
wedges 47
welding 181
 plastics 217
wheel
 balancing 146
 dressing 145
workholding 126, 143, 163

zinc
 alloys 273
 coating 208